University of Lowell
Libraries
UNIVERSITY of LOWELL
ALUMNI/LYDON LIBRARY
AF704204

SYNTHETIC POLYMERS

and the Paper Industry

Also published by Miller Freeman Publications for the pulp and paper industry:

Manual of Supercalender Operations
by Francis E. Schiller

Metrication for the Pulp and Paper Industry
by Kenneth E. Lowe

An Introduction to Paper Industry Instrumentation
by John R. Lavigne

Practical Computer Applications for the Pulp and Paper Industry
edited by Kenneth E. Lowe

Fiber Conservation and Utilization: Proceedings of the May 1974 Pulp & Paper Seminar, Chicago, Illinois
edited by Paul D. Van Derveer and Kenneth E. Lowe

Transport and Handling in the Pulp and Paper Industry Volume I: Proceedings of the first international symposium sponsored by Pulp & Paper International, Rotterdam, Netherlands, April 1974
edited by Leonard E. Haas and John E. Kalish

Transport and Handling in the Pulp and Paper Industry Volume 2: Proceedings of the second international symposium sponsored by Pulp & Paper International, Rotterdam, Netherlands, October 1976
edited by Leonard E. Haas and John E. Kalish

International Glossary of Technical Terms for the Pulp and Paper Industry
edited by Paul D. Van Derveer and Leonard E. Haas

Atlas of the Western European Pulp and Paper Industry: showing all pulp, paper, and paperboard mills in Western Europe with over 5,000 tons capacity

SYNTHETIC POLYMERS and the Paper Industry

VLADIMIR M. WOLPERT

A PULP & PAPER BOOK

MILLER FREEMAN PUBLICATIONS, INC.

TS 1124 .W64
Wolpert, Vladimir M.
Synthetic polymers and the
paper industry

Copyright © 1977 by Vladimir M. Wolpert

Library of Congress Catalog Card Number: 76-6719
International Standard Book Number: 0-87930-056-6

All rights reserved. No part of this book covered by the copyrights hereon may by reproduced or copied in any manner whatsoever without written permission except in the case of brief quotations embodied in articles and reviews. For information contact the publishers, Miller Freeman Publications, Inc., 500 Howard Street, San Francisco, California 94105, USA.

Printed in the United States of America.

CONTENTS

2. *Definition of Different Types of Synthetic and Plastics Papers* *53*

3. *Synthetic Papers* *56*

4. *Synthetic Pulp Processes and Feedstocks* *123*

LIST OF TABLES

CHAPTER 1

CHAPTER 3

CHAPTER 4

CHAPTER 5

CHAPTER 6

PREFACE

This book describes the manifold relations between the traditional pulp/paper and the synthetic polymer industries, and has been written to assist those in the paper industry who need to know where their industry stands today with respect to synthetic polymers. For those who are working in the synthetic polymer industry, the book will help in their activities by providing them with an understanding of the pulp/paper industry. The author feels strongly that those working in these two industries would greatly benefit by a better understanding of the interrelations between them. In addition the book will be useful to executives of firms which supply equipment and materials to the pulp/paper and the plastics industries, as well as to converters of paper, paperboard, and plastics materials, since it provides information on the marketing trends of these products.

For more than six years the author has been collecting the information that makes up this book. He has repeatedly visited industrial firms and research institutes in various countries, and has taken part in international exhibitions and conferences, where he has presented papers on this subject and has exchanged views with many people in both of these industries.

Cooperation and competition between the polymer industry and the pulp/paper industry are described in the first chapter of this book and are referred to throughout the other chapters. Special chapters describe the development of synthetic paper, synthetic pulp, and plastics paper, with technical data on individual materials and products supplied by various manufacturers. Other chapters deal with the definition of different types of synthetic and plastics papers (about which there is still come confusion in many quarters), the impact of oil prices, environmental impact, price developments since September 1973, and the outlook for both industries. The author, who has held seminars at R&D departments of manufacturing enterprises and at universities, feels that university students—the industry managers of tomorrow—will be well advised to come to grips with these problems today.

Every effort has been made to incorporate in this book the latest available statistical data. However, the recession that began toward the end of 1974 and continued through 1975 has in some cases distorted the overall trend. In these cases, data from previous years have been used.

V.M.W.
London

INTRODUCTION

In describing the manifold relations that exist between the traditional pulp/paper and the synthetic polymer industries, this book stresses that cooperation and competition between the two industries will be intensified in the years to come, and that more firms will be engaged in activities involving both industries.

The synthetic polymer industry is both younger and smaller (tonnagewise) than the pulp/paper industry; the major plastics materials (PVC, PS, LD PE) were not introduced on an industrial scale until after 1930. And, in fact, plastics materials such as high-density polyethylene (HD PE) and polypropylene (PP) became available only about 20 years ago. Since then, the plastics industry has developed products for a wide range of applications. Some of these products are finding an increasing market before, during, and after the actual papermaking process, improving the finished products as well as the production processes of the paper industry. There is thus a growing trend within the paper industry to develop from a purely mechanical industry toward a mechanical and chemical industry, which represents one aspect of the cooperation between the two industries.

The trend of increasing production of chemical by-products in the pulp/paper industry (lignosulfonates, tall oil, crude turpentine, yeast, proteins, etc.) is another aspect of increasing cooperation between the pulp/paper industry and the chemical industry, of which the synthetic polymer industry is a part.

New products have been developed from the raw materials of both industries, with functional properties that could not be achieved by using materials from only one industry. This trend can be witnessed in many other industries too, and reflects the modern marketing approach, which starts from the end product and its required properties, and then evaluates the most appropriate raw materials and processes to be used for making this product.

Further evidence of growing cooperation between the paper industry and the synthetic polymer industry can be seen from the leading paper-making and paper converting firms diversifying their manufacturing activities into the field of plastics, in some cases replacing paper with plastics. These are cases of cooperation overlapping into competition from the point of view of the use of raw materials. In the packaging field, an important outlet for both industries, materials and products made from both paper and plastics have been developed; at the same time, plastics have been competing with and replacing paper and board in some applications. While the competition is not new, the latest developments are especially significant because certain plastics films are for the first time being called *plastics paper*, and this material has been securing a growing share of the wrapping and retail-bag markets.

The development of various types of synthetic paper for the high quality printing- and writing-paper market has shown, with a few notable exceptions, that these new products have been too expensive—their production has therefore been discontinued. However, the experience gained in the development of these products enables the manufacturers to proceed with their R & D work in order to bring out synthetic paper products of a second generation. Some spunbonded materials show increasing sales in the paper industry's markets.

The development of synthetic pulp can be described as the introduction of a new, additional raw material for the paper industry, starting as an "additive" to traditional pulp, and processed on conventional papermaking equipment in order to produce paper with superior properties for special applications. Synthetic pulp can be considered as a further aspect of cooperation between the paper and the synthetic polymer industries. Synthetic pulp also has a potential market as binders in nonwoven production and in applications outside the paper industry.

The increased price of crude oil has influenced the production costs of synthetic polymers. In finished plastics products, however, this cost factor is diluted to a large extent. The production costs of other materials, including paper, have gone up because of increased energy costs resulting from the higher price of crude oil.

Energy consumption in the production of plastics on a weight basis is lower than that of many other materials, and the comparatively low specific weight of plastics must also be taken into consideration when evaluating the energy consumption on an area basis of the material in question. The competitive strength of plastics has remained unchanged compared with the situation prior to the crude oil price increases in September 1973.

Since the end of 1973, prices of paper and board have registered significant increases from their very low, partly uneconomical level. During the recession that started in the second half of 1974, the trend of price increases was interrupted, but this was a temporary situation only. Within the overall trend of the growing consumption of paper and board and the required financial returns of the industry enabling the securing of capital for investments in new production units, the prices of paper and board will have to increase at a rate higher than that of the overall inflation if the growing demand is to be satisfied.

Facing page: scanning electron micrograph (magnification: 3000x) of a sheet made with 100% SWP® synthetic pulp. Courtesy Crown Zellerbach.

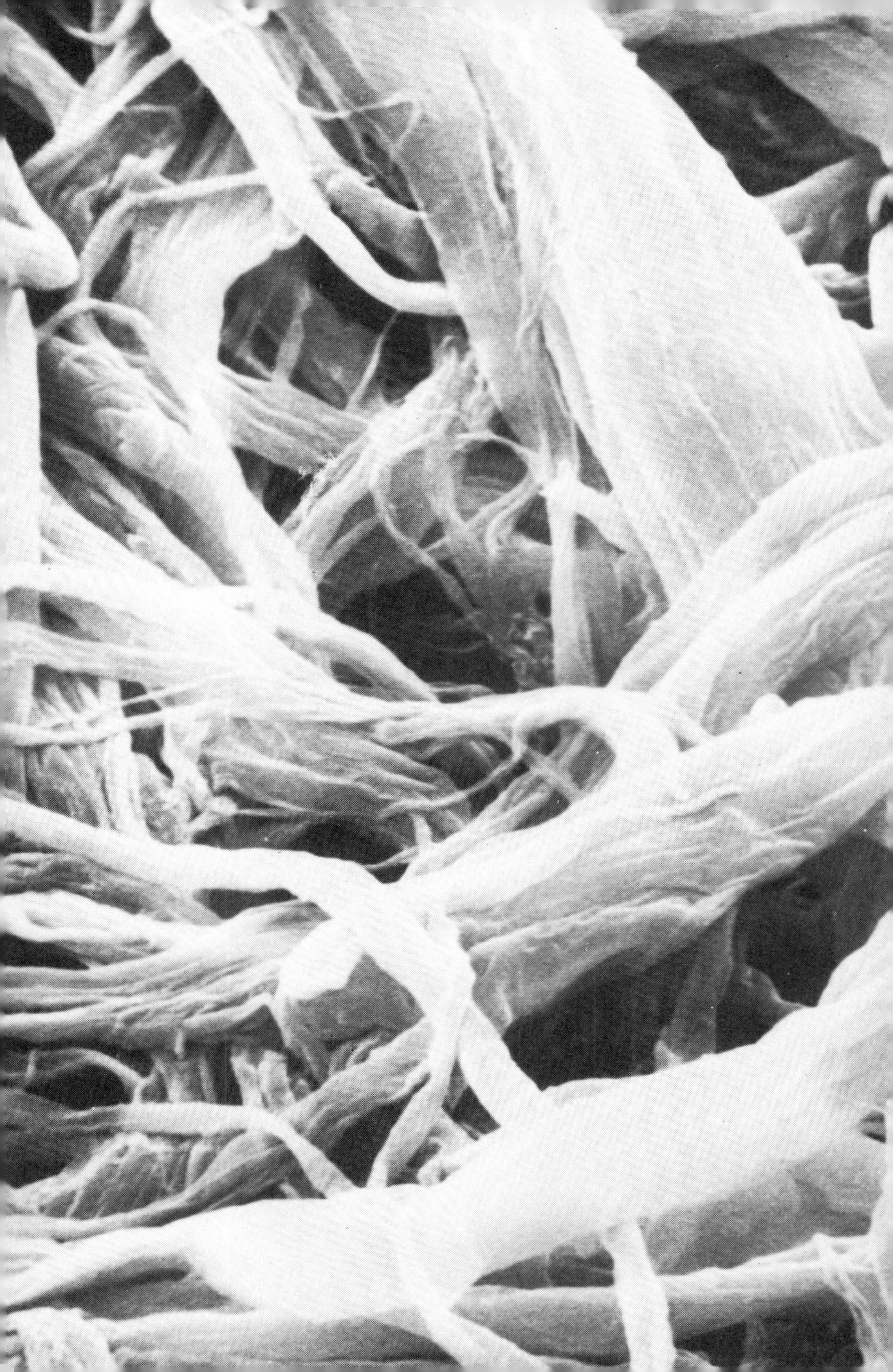

ABBREVIATIONS

LD PE	low-density polyethylene
HD PE	high-density polyethylene
PP	polypropylene
OPP	oriented polypropylene film
PS	polystyrene
XPS	expanded polystyrene
PVC	polyvinyl chloride
PVdC	polyvinylidene chloride
EVA	ethylene-vinyl acetate copolymer
MT	metric ton
TPA	tons per year
MD	machine direction
TD	transverse direction

1. Cooperation and Competition between the Polymer Industry and the Pulp/Paper Industry

Cooperation between the pulp/paper and board industry and the synthetic polymer industry, including plastics, is showing an upward trend. More and more synthetic polymer products are being used before, during, and after the actual papermaking process. New synthetic polymer products have been (and are being) developed for the paper industry in order to improve the productivity of papermaking and to create paper products possessing better functional properties. In addition, when synthetic polymer products are used as retention aids they contribute to cost savings by reducing losses of mineral fillers and fine fibers; they also reduce stream pollution by paper mill effluents.

The range of laminates and coated materials, consisting of paper and plastics materials, is increasing steadily and is creating new outlets for products of the paper industry. These new products find application mainly in the packaging field, in some instances replacing previously used products made of glass, metals, and textiles. And it is in the packaging field, as well as in some sectors of the engineering industries, that *competition* by plastics against paper and board has been particularly strong and is continuing to increase.

The paper and board industry worldwide has a total annual production of about 140 million metric tons (MT), and all indications show that the steady increase in paper and board consumption will continue. Population growth and the rising standard of living are the main factors contributing to this trend.

The plastics materials industry, a much younger industry than the paper industry, amounts to an output of more than 40 million MT per year. This industry supplies a wide range of different markets, and it is estimated that about 25% of the total tonnage (about 10 million MT) is used in packaging. Of this, a large share has penetrated the paper and board markets of packaging applications.

Table 1.1 *OECD Production*

	1972		*1974*	
	Tonnage (million MT)	*Approximate share of overall output (%)*	*Tonnage (million MT)*	*Approximate share of overall output (%)*
Newsprint	18.4	16	19.4	16
Other printing and writing paper	24.8	22	28.6	23
Wrapping and packaging paper and board	52.3	46	55.8	45
Construction paper and board	2.2	2	2.3	2
Household and sanitary paper; special thin paper	7.0	6	7.5	6
Other paper and paper board	8.6	7	9.6	8
Total	113.2	100	123.2	100

Source: 1972 data compiled from OECD, *The Pulp and Paper Industry 1972-1973*, Paris, 1974; 1974 data compiled from OECD, *Pulp and Paper*, "Quarterly Statistics," Issue 1975/4, Paris, 1976.

Table 1.1 analyzes paper and paperboard production by main groups of products. Based on 1972 and 1974 output in OECD countries[1] (which accounts for nearly 85% of the world output), the table shows the individual shares of these products in the total production tonnage of the industry.

COMPARISON OF PLASTICS CONSUMPTION WITH PAPER AND BOARD

In comparing the consumption of plastics with that of paper and board, it is important to take into account the considerably lower specific weight of plastics. In the replacement of paper by plastics (and in comparing their respective prices) the comparison on an area basis provides a better appreciation than on a weight basis.

In fact, it can be assumed that depending on the specific products, 1 ton of plastics products replaces between 1.6 and 2 tons of paper

1. The Members of OECD are Australia, Austria, Belgium, Canada, Denmark, Fed Rep Germany, Finland, France, Greece, Iceland, Ireland, Italy, Japan, Luxembourg, Holland, New Zealand, Norway, Portugal, Spain, Sweden, Switzerland, Turkey, United Kingdom, USA.

and board. The development of *oriented* (stretched in one or both directions) plastics films and of *expanded* (foamed) plastics sheets results in a savings of plastics materials per area unit and increases the ratio of replaced weight of paper and board by plastics per weight and area units.

Paper and Board in Packaging

Wrapping and packaging represents the largest single outlet of the paper and board industry. In 1974, wrapping and packaging paper and board production and its share of the total production of the paper and board industry, broken down by the main OECD regions, was as shown in Table 1.2.

Plastics in Packaging

Packaging is also the largest single outlet for plastics. It is estimated that packaging consumption accounts for from about 20% to 30% of the total plastics-materials consumption, varying within this range from country to country. However, in contrast to the paper industry, where statistics are available for wrapping and packaging paper and board, in the plastics industry the difficulty of establishing the exact tonnages used in packaging is due to the fact that the same polymers find applications in different markets (plastics films are used for packaging and also for agricultural and other purposes). In addition, packaging materials and finished packages from plastics are made by independent converters, as well as by industrial firms that use plastics both for production of their goods and for in-house manufacturing of packages for their own large-scale use.

Table 1.2 *Wrapping and Packaging Paper and Board Production (1974)*

	Production (million MT)	*Approximate share of overall output (%)*
Western Europe	17.2	42
Canada	2.6	20
USA	27.5	52
Japan	7.5	48
Australia and New Zealand	1.0	62
Total	55.8	45

Source: Production data compiled from OECD, *Pulp and Paper*, "Quarterly Statistics," Issue 1975/4, Paris, 1976.

In the USA it is estimated that in 1974 more than 2.7 million MT of plastics, including more than 1 million MT of films, were used in packaging, accounting for over 22% of the total plastics-materials consumption. It is expected that packaging will continue to be among the high-growth areas of plastics (including for films, merchant bags, grocery bags, etc.).

In France, the 1974 consumption of plastics in packaging was estimated at nearly 500,000 MT, or about 22.5% of the total plastics consumption; LD PE film accounted for about half of the tonnage of plastics in packaging.

In Britain, about 600,000 MT of plastics, including about 315,000 MT of films, were used in packaging in 1974, accounting for about one-third of the total plastics consumption. The food industry uses approximately half of the plastics packaging total.

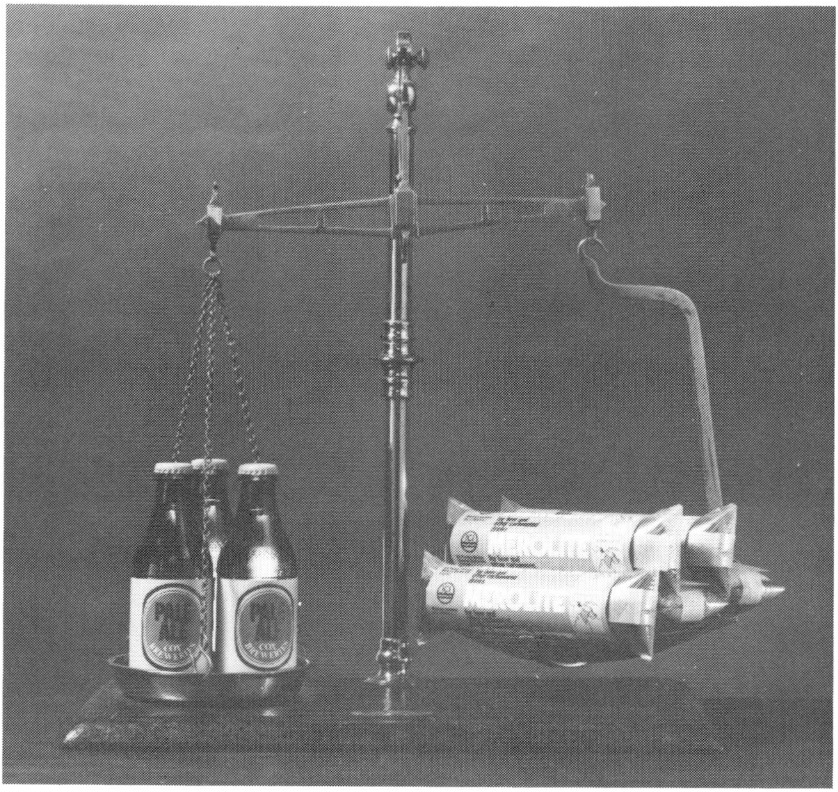

Plastics are not always direct competition to paper. Here plastic combines with paper to produce a new pack for beer and carbonated drinks, competing with glass. Courtesy ICI Plastics Division.

Plastics compete in packaging with glass, metal, textiles, and paper and board. At the same time, in some applications plastics *together* with paper have been successfully replacing jute and other textile sacks, and in other applications plastics *together* with paper have been competing with glass, thus providing new outlets for the paper industry. Merolite, a new pack for beer and other carbonated drinks, developed by ICI (England), consists of a tubular pouch of coated polyester film with a paper sleeve. The Merolite Pack currently in production has a capacity of 25 cl (8.8 fl oz) and, at 10 grams (g) (⅓ oz) is lighter than any other pack used for beer and carbonated drinks.[2]

At present the ratio of paper and board production to the plastics material production on a worldwide basis tonnagewise is about 3.5:1. However, the rate of growth of plastics has been considerably higher than that of paper. The world production of plastics materials increased from 6.9 MT in 1960 to 14.0 million in 1965 to 29 million in 1970, thus more than doubling every five years during that period. Since then, the world output (in million MT) is estimated to have increased as follows:

Year	*Output*
1971	31
1972	37
1973	40
1974	41

Thus between 1971 and 1974 plastics production increased 32.3%.

Packaging is the biggest single market of the paper and paperboard industry; on a worldwide basis about 60-65 million MT of paper and paperboard are used annually in this market.

Plastics are supplied to a wide range of user-industries, and packaging accounts for about 20-30% of the overall plastics consumption (the situation varies from country to country). This means that at present about 10 million MT of plastics find application in packaging. The U.S. plastics materials industry, the world's largest polymer industry, had an output of 24,967 million pounds (lb) in 1974 (over 11.2 million MT), registering an increase of 2.4% over the 1973 production. The ouput of all thermoplastics increased by 3.8%, reaching 21,473 million lb in 1974 (the output of the main thermoplastics is given in Table 1.3). HD PE showed the greatest rate of increase. It is estimated that in 1974 about 2.5 million MT of plastics were used in packaging (about 22% of the total plastics materials production).

2. ICI Plastics Division "Technical Note No. 101," Welwyn Garden City, England, May 1973.

Table 1.3 *U.S. Output of Main Thermoplastics*

	1974 Output		*Increase over 1973*
	(Million lb)	*(Million MT)*	*(%)*
LD PE	5,973	2.7	2.9
HD PE	2,837	1.3	7.6
PP polymers & copolymers	2,249	1.0	4.0
Styrene resins (solid)	3,364	1.5	1.3
PVC & copolymers (dry resin content)	4,850	2.2	6.3

The 1973 output of plastics materials in Japan amounted to nearly 6.5 million MT. In Europe, the 1973 production of plastics is estimated to have amounted to about 21.5 million MT, including 17.0 million in West Europe and about 4.5 million in the Soviet Union plus other East European countries. In Fed Rep Germany, the biggest European manufacturing country of plastics materials, the output decreased from 6,430,000 MT in 1973 to 6,380,000 in 1974, and still the 1974 production was nearly 31.5% higher than the 1971 production. The per capita consumption of plastics in Fed Rep Germany increased from 62.5 kilograms (kg) in 1971 to 82.4 in 1973, but dropped to 76.9 in 1974.

In Fed Rep Germany, the 1972 total packaging materials consumption was valued at about DM13,000 million; of this total, paper plus board accounted for 44.3%, plastics for 22.3%, metal for 20.6%, and glass, wood, cork, etc., for the remaining 12.8%.

Plastics Consumption Forecast

In 1971, several forecasts were made envisaging the world annual output of plastics reaching 90 million MT by 1980. According to these forecasts, the production breakdown (in million MT) by main industrialized areas was expected to increase as follows:

	1970	*1980*
Western Europe	11.0	30
USA	8.5	20-30
Japan	5.0	15

It appears, however, that though these projected production figures will be reached, they will probably be delayed by a few years. The output of *packaging films* by main types (world total) was forecast to increase as is shown in Table 1.4.

Table 1.4 *Packaging Films by Main Types – World Total*

	1968 (Actual)		*1968-75 (Forecast)*	*1975 (Forecast)*		*1975-85 (Forecast)*	*1985 (Forecast)*	
	Total (thousand MT)	*Share (%)*	*Average growth rate (%)*	*Total (thousand MT)*	*Share (%)*	*Average growth rate (%)*	*Total (thousand MT)*	*Share (%)*
Polyethylene LD	1,300	59	14	3,250	65	11	9,400	72
Polyethylene HD	50	2	26	250	5	10	650	5
Regenerated cellulose	400	18	–	400	8	–4.5	250	2
PVC	250	11	12	550	11	6	1,000	8
Polypropylene	100	5	17	300	6	10	800	6
Others	100	5	14	250	5	14	900	7
Total	2,200	100	12.5	5,000	100	10	13,000	100

Source: J.F.L. Whitfield, ICI Plastics Division, "Packaging Films in the Future" (Presentation at The Plastics Institute Conference, "Developments in Films for Packaging," London, February 1972), quoted in Wolpert, *Synthetic and Plastics Paper*, Vol. 3, 1972, p. 378.

Note: The repercussions of the crude oil price increases at the end of 1973/beginning of 1974 on the competitive strength of plastics in replacement of paper and board are described in Chapter 6 of this book.

According the the estimates of the German Plastics Manufacturing Association, the consumption of plastics in the European Economic Community is expected to reach about 24 million MT/yr by 1980, including 7 million MT in Fed Rep Germany (as against 4.3 MT in 1972) and about 4 million MT in Britain.

The Stanford Reseach Institute (SRI) has forecast that the yearly production of plastics materials (resins) in the USA will increase from 7.6 million MT (16,730 million lb) in 1970, "most probably" to 103 million MT (226,600 million lb) by the year 2000 (adding that the yearly total may range from 77 to 129 million MT), and that the end use of plastics in packaging will increase from 2,040,000 MT in 1970 to 21.5 million MT by the year 2000, an annual average growth rate of 8.2%.[3]

The SRI report stated that "some areas of significant growth are bottles for foods, shrink wrap, paper-like films, pouches for liquid packaging, plastic coated glass bottles, plastic lids, oil cans, plastic strapping, shipping bags, refuse bags, pallet wraps, shipping pails, and plastic milk bottles. Possible new uses for plastics in packaging are aerosols, carbonated shipping boxes, paint cans, and wine bottles." This report, which was based on a forecast compound growth rate of 9.1% for total plastics production and 8.2% for plastics in packaging, was issued before the increase in crude oil prices that occurred in September 1973 and before the recession that began in the second half of 1974.

Early in 1975 leading U.S. plastics manufacturers estimated that whereas the total plastics production would show a decrease in 1975 compared with that of 1974, starting with 1976 the U.S. plastics industry's production would register a compound growth rate of about 8% per year. It is significant that leading polymer manufacturers have been investing at a high level during the recession (counter-cyclic investment policy) in extension of their already existing and their new production units. Leading oil companies have been extending their downstream activity into the production of monomers and polymers, and even into the production of plastics goods.[4]

The latest estimates and the plastics industry's investment decisions clearly indicate that the industry is confident of future plastics-consumption growth. Even if the SRI forecast of a "most probable" 103 million MT/yr consumption in the USA by the year 2000 is considered too optimistic, the lower figure of this forecast at 77 million MT/yr seems to be realistic.

3. SRI, "The Plastics Industry in the Year 2000" (Report prepared for The Society of the Plastics Industry, Inc., and The Society of Plastics Engineers, Greenwich, Connecticut, USA), Menlo Park, California, April, 1973.

4. Vladimir M. Wolpert, *Plastics and the Paper Industry* (London: Wolpert & Jones Studies Ltd., 1975), pp. 16-23.

In Western Europe it is expected that from 1976 onward the consumption of plastics will again show a high growth rate and that the amount of 30 million MT/yr (mentioned in the 1971 forecasts) will be reached by 1980, or one or two years later.

During the recession (end of 1974-75), the packaging industry proved to be more resilient than some other industries, including the motorcar and building industries, on both sides of the Atlantic.

In the USA, according to a report by Arthur D. Little, Inc., the use of plastics in packaging (in million lb) increased as follows:[5]

Year	*Use*
1958	700
1967	2,600
1970	3,925
1972	nearly 5,000
1974	about 5,000

The annual growth of plastics in packaging was about 15% from 1965 to 1970 (on a weight basis), and it was estimated that the annual growth of plastics "should average about 11% per year through 1975."[6] The projected growth rate was not achieved during the latter period, first because of the shortage of plastics immediately before the recession, and then because of the recession.

Therefore the forecasts in this report regarding the volume of plastics in packaging (nearly 6.5 billion lb by 1975 and nearly 10.2 billion lb by 1980) seem to be overly optimistic. The consumption of plastics in packaging wil be nearer to 8 billion lb annually by 1980.

In 1971, plastic packaging was valued at $2.1 billion and accounted for 11% of all packaging materials used in the USA that year, up from 6% in 1960.[7] The discrepancy between the growth of plastics in volume and value during that period is the result of a decrease in plastics prices, which resulted from technological developments and economy of scale in the production of plastics. In the industrialized countries of Western Europe the share of plastics in packaging, that is, of the total tonnage of all materials used in packaging, is higher than in the USA, and the penetration of the packaging market by plastics is expected to continue at a high rate.

5. Arthur D. Little, Inc., "Packaging in Perspective" (Report to the Ad Hoc Committee on Packaging), February 1974.

6. Ibid., pp. 25 and 52.

7. Ibid., p. 52.

M. Pariat, reviewing the recent study *Packaging 1980* by the Battelle Institute of Geneva, writes that "the production of low-density polyethylene (LD PE) small and medium-sized bags is expected to double, that of heavy bags to treble in the Federal Republic by 1980, while the French foresee fivefold growth in the production of heavy duty plastics bags. Paper, they think, will be more popular for automatic wrapping . . . in the UK it is expected that the use of paper will remain fairly constant, while that of plastics will double."[8]

According to Pariat the production of small and medium-sized paper bags in Fed Rep Germany decreased from 166,000 MT in 1973 to 153,000 MT in 1974, whereas the production of these bags from plastics increased from 235,000 MT in 1973 to 262,000 MT in 1974. In the case of heavy-duty bags the West German production of paper bags increased from 264,000 MT in 1973 to 275,000 MT in 1974, and the production of plastics bags increased from 80,000 MT in 1973 to 85,000 MT in 1974.

TECHNOLOGICAL COOPERATION

The technological cooperation between the pulp/paper and synthetic polymer industries is reflected in:

1. The growing use of synthetic polymers by the paper industry.
2. The development of applications of plastics films in combination with paper and board, in this way opening new markets for both industries. New packaging systems combining shrink film and board trays are an example of this type of development.[9]

The growing use of synthetic polymers by the paper industry indicates the trend of papermaking developing from a purely mechanical process to a mechanical and chemical technology, resulting in products with better functional properties. Dr. A. A. Robertson, of the Pulp and Paper Research Institute of Canada, referred to

> The encouraging possibilities of the continued development of products based on combinations of synthetic polymers and paper. Incorporation of synthetic polymers into the sheet are effected by additions of synthetic polymers *before, during and after* the papermaking process, whereby the synthetic polymers are added as

8. Maurice Pariat, "Precarious Prospects for the Paper Bag," *Pulp & Paper International*, 17, no. 11 (October 1975), p. 56.

9. See Vladimir Wolpert, *Shrink Film—A Growing Market* (London: Wolpert & Jones Studies Ltd., 1973).

solution or latex or as synthetic fibers. As a result of this addition, the mechanical properties of the sheet are altered and reflect the properties, distribution and interactions of the polymer (italics mine).[10]

Dr. Robertson discussed the incorporation of polymers into paper using the following three methods:

1. Solution impregnation of a bulky saturating paper with polymer in an organic solvent.
2. Emulsion or latex impregnation of a similar formed sheet.
3. Addition of latex to stock before sheet formation.

The importance of the paper industry as a market for the chemical industry has been referred to in the annual reports of some leading chemical companies. For example, in its 1972 Annual Report, Hercules Inc. listed the breakdown of its commercial sales to principal user-industries. The report showed that the paper industry's share accounted for 10% of the overall sales of the company in 1971 and 1972, and it stated that the upward trend in paper and paperboard production continued through 1972 and had an effect on the paper chemicals business that was quite favorable. " 'Neuphor' the base product for a new papersizing system showed dramatic growth, and additional capacity is being planned."

Monsanto Company (USA) said in its 1972 Annual Report that the growth of the paper industry during 1972 led to increased sales of paper chemicals. (Paper chemicals are used to impart wet strength and water resistance to paper products.) A new wet-strength resin, Santo-Res 23, was introduced. The report also referred to plastics pigments based on polystyrene particles being tested to replace inorganic pigments. In its 1973 Annual Report, Monsanto referred to the strong demand for paper chemicals: "Paper chemical sales paralleled paper industry performance. Recognized technical leadership in surface sizing expanded business opportunities."

It is significant that Borden Chemical (Division of Borden, Inc.), the beginning of which was the acquisition of the Casein Company of America (which is claimed to be the largest casein supplier in the USA) has diversified into the field of synthetic polymers and is now supplying, in addition to casein and modified casein products, a wide range of

10. A. A. Robertson, "Modification of the Mechanical Properties of Paper by the Addition of Synthetic Polymers" (Paper presented at the International Fundamental Research Symposium, "The Fundamental Properties of Paper Related to its Uses," held in Cambridge, England, in September 1973 by the Technical Section of The British Paper and Board Makers Association).

chemicals for the paper industry, including binders, beater additives, PVC latices and polyvinyl alcohol, wet strength resins, etc.

BASF (Fed Rep Germany) referred in its 1972 Annual Report to the new surface sizing agents which "have proved their advantages." The report also referred to the various BASF dispersions. BASF has been very active in developing and manufacturing chemicals for the paper industry and has in its Ludwigshafen plant a full-scale department with up-to-date machinery for the paper industry. It has published the *Manual for the use of BASF products in the paper industry* (167 pp.), which describes the wide range of products for the paper industry made available by the company.

According to some estimates, in the USA about 90% of all lightweight coated (LWC) paper is printed by offset and letterpress processes, whereas in Western Europe more than 60% of LWC paper is rotogravure printed.[11] This has been possible because of the introduction of special binder systems (e.g., acrylic sole binder system).

In Fed Rep Germany, pigment-coated paperboard and pigment-coated paper have registered a higher growth rate than the overall production increase of paperboard and paper.

AUXILIARIES FOR PAPER IN WESTERN EUROPE

D. A. Clark, of English China Clays, has estimated that the paper industry in Western Europe uses about 4.5 million MT/yr of auxiliary products, valued at approximately £200 million. This total can be subdivided as follows:[12]

about 3.25 million MT mineral pigments
about 0.5 million MT bonding agents, including coating binders
about 0.5 million MT sizing materials
about 0.25 million MT various products, including retention aids, wet strength additives, effluent treatment products, dyestuffs, etc.

Within the overall market of auxiliaries for papermaking, the penetration by synthetics varies from sector to sector. Clark made the following points:

1. An almost complete replacement of natural products by synthetics has taken place in the sectors of dyestuffs, bleaching agents, wet

11. Vladimir Wolpert, *Synthetic and Plastics Paper* (London: Wolpert & Jones Studies Ltd., 1971-) Vol. 4, 1974, p. 726.

12. Presentation at the European Chemical Market Research Association Conference, October 1973.

strength resins, and coating lubricants.

2. Synthetics account for over 50% of the total market of retention aids and dispersants.
3. In the sector of coating binders (the overall market has trebled to about 300,000 MT/yr within the last 10 years) synthetics have achieved a considerable market penetration, and synthetic latices are reported to account for about two-thirds of the total market. Whereas the share of casein has dropped considerably, mainly due to strong price fluctuations and availability of this natural product, starch, another natural product, has been keeping its market share. (The 1972 Annual Report of Feldmühle AG says that the sustained price increase for casein, which is used as a binding agent for coating colors, was partially countered by the alternative use of proteins and synthetic binding agents.)
4. Synthetics have achieved only a slow penetration of the sizing agents market. However, it is expected that the present synthetic share of 5% of the market will increase in the future.
5. The biggest sector of auxiliary products is accounted for by mineral pigments used for filling and coating. In Western Europe this market sector is dominated by three basic minerals:

   a) China clay, which accounts for 80% of the total filler market and for over 90% of the coating market.
   b) Calcium carbonate or chalk, which accounts for 5% each of the filler and the coating market.
   c) Talc, which accounts for about 10% of the filler market.

   The penetration of this low-cost price market has been insignificant (either in small specialty applications or as minor additives to mineral pigments).

In regard to the total market of auxiliary products for the paper industry, synthetic materials account at present for about 400,000 MT/yr, or about 10% of the total auxiliaries market of 4.5 million MT. However, if one excludes the large volume of low-cost fillers and coating pigments, synthetics account for about one-third of the remaining market. It must be noted, however, that if the synthetics share is measured in monetary terms (i.e., in value and not in volume), then the above mentioned figures undergo considerable change in favor of synthetic materials.

These figures refer to the use of synthetic polymer products only as auxiliaries in paper production; they do not include the increasing tonnages of plastics used in the coating of paper and board after the actual papermaking.

NONWOVENS

The production of nonwoven fabrics represents a further area of cooperation between the paper and man-made polymers industries since a number of modern machines can be used for the production of both paper and nonwovens (applying the wet-lay process).[13] These machines can process hydrophilic natural fibers (e.g., cellulose) of comparatively low length to paper, as well as hydrophobic man-made polymer fibers to nonwoven fabrics, mainly for textile uses.

The decision whether to produce paper or nonwoven fabrics on this equipment depends on economic considerations, such as marketing potential and profitability. Papermaking firms that have no organizational connections with pulp mills are apt to be less confined in the choice of raw materials than the papermaking division of a vertically integrated concern, which may prefer to use pulp supplied by the pulp mills of the same concern. However, with the growing tendency of accountability of individual divisions within big industrial concerns it is evident that higher rates of plant utilization and profitability are becoming the decisive factors in the choice of products to be made and raw materials to be used.

ORGANIZATIONAL COOPERATION

One aspect of organizational cooperation is the trend of papermaking firms to diversify their activities into the field of the plastics industry. A large number of papermaking firms in various countries have diversified their activities into fields outside papermaking. This is because the financial returns on actual papermaking have been very low and the cycle of boom/recession in paper trade has been shortened.

It is significant that the two leading British paper concerns—Bowater and Reed—have recently dropped the word *paper* from their names, and that Bowater was advertising the fact that more than 50% of its total sales was coming from nonpaper activities.

The diversification into the field of plastics has various reasons, including the high growth rate of this industry and the penetration by plastics into markets that have been supplied by the paper and board industry (i.e., packaging and building).

St. Regis Paper Company, a vertically integrated concern with its own timberland and plants, produces finished paper and board products. In 1972 the company had a turnover of more than $1,000 million. In its

13. The spunbonding process based on the utilization of a different type of machine is described in Chapter 3.

1972 Annual Report, the company referred to its diversification into plastics by saying:

> Among the most far-reaching changes in recent years in the broad field of bag and flexible packaging served by St. Regis has been the emergence of plastics as a dominant material. St. Regis, instead of seeking to shelter its paper-related packaging markets from plastics, has sought every way to bring the benefits of the material to its packaging customers. It has been an innovator in developing plastic films, paper-plastic combinations, plastic laminations, and plastic coatings. The Flexible Packaging Division has gradually converted almost all its breadwrap volume from waxed paper to plastic bags over the last decade and has gained a pre-eminent position in the technology of plastic coating and laminating. This has been important in establishing the division as a leading supplier to the snack-food and convenience-food industries.
>
> The Bag Packaging Division, once totally in paper, now derives 35 percent of its sales from packaging that is part plastic or all plastic, and has fast-growing business in such areas as shrink-film. Our other packaging and converting divisions have also moved heavily into plastic, and St. Regis aims for continued growth in this important field.

In its 1974 Annual Report St. Regis said that "It was a good year to be in the packaging business in 1974 . . . ", and the company estimated that in the USA it had a 5.5% share of a total packaging market of $12 billion. Its profit from packaging and converted products operations accounted for 42% of the total profits in 1974 as against 33% in 1973.

The section of the report dealing with the Bag Packaging Division referred to the fact that this division offers a wide range of paper and plastic packaging—together with the systems to use it—and that the demand was brisk across the full line in 1974. The report said that "plastic sheeting and shrink film made major gains" and also that "cross-plastic bags really came into their own as more raw material became available . . ."

Crown Zellerbach stated in its 1974 Annual Report that "in the case of Flexible Packaging, demand continued particularly high for the special plastic materials and constructions developed mainly through our own research and used primarily in the processed food industry." The report also stated that the company's small Rigid Plastics Division posted its most successful year.

R. A. Oswick, managing director of the papermaking concern Reed & Smith Ltd., England, chairman of the Association of Packaging Papers, and vice-chairman of the European association Embalpack, expressed the view that in packaging "There will be some areas where plastics will take the major part of the growth, and there is certainly nothing to be gained by the manufacturers of one product trying to knock the other product." He felt that there is going to be (and is already) scope for the development of usage of both paper and plastics in plastics laminated paper. He referred to some packaging paper types which have been hit by plastics films, and he stressed the point that in some applications it is difficult to deny that the plastics material is more acceptable.[14]

The above cited randomly selected statements by executives of papermaking concerns from both sides of the Atlantic indicate increasing *organizational cooperation* between firms of the pulp/paper and the synthetic polymer industries.

The advent of new synthetic products for the graphic industry has resulted in papermaking firms marketing these products manufactured by firms of the synthetic polymer industry (e.g., synthetic papers and spunbonded materials) to the graphic industry, as well as in the establishment of joint enterprises by firms of both industries to manufacture synthetic papers and synthetic pulp.

For over 25 years Kimberly-Clark Corporation (USA) has been manufacturing and marketing "durable printing materials," including Texoprint plastic paper, and latex-saturated, fiber-based label papers. According to the Texoprint brochure, Texoprint is a durable plastics printing paper, consisting of a balanced combination of strong bleached kraft fibers and latex plus opacifiers and printing fillers. The brochure states that

> Bleached Kraft fibres are formed on the Fourdrinier into a low density absorbent sheet which is then saturated with elastomeric polymer latex and opacifying fillers (picking up 60 percent of its own weight in dry solids). The then dried sheet is coated on both sides with printing fillers (including TiO_2) and a flexible latex binder (made from special, non-specified "plastics").

It is stressed that latex replaces the normal cellulose bonding material in the fiber base, and that in the coating, flexible synthetic resins replace conventional starches, glue, or casein binders. Whereas Kimberly-Clark describes Texoprint as a plastics paper, in accordance with the classifications developed in this book this product belongs to the category of synthetic paper.

Kimberly-Clark has also been manufacturing Spun-Tuff, a spunbonded polyethylene sheet Tyvek (made by Du Pont), coated on one side in order to achieve improved printing quality. Spun-Tuff and

14. *Packaging Review*, London, March 1973.

Texoprint are used for the same applications, including cover mounts, flip charts, games, pressure sensitive and other labels, wall charts, manuals, signs and posters, etc.

In May 1973 Kimberly-Clark and Oji-Yuka Synthetic Paper Co. of Japan announced the conclusion of an agreement that gives Kimberly-Clark the exclusive sales rights in North and Central America for Oji-Yuka polypropylene "durable materials" (synthetic paper). This contract included an option for a joint venture or a licensing agreement for manufacturing this synthetic paper in the USA. Kimberly-Clark has been testing the market and evaluating this synthetic paper for printing, converting, and packaging applications. In the USA this product has been marketed under the trade name Kimdura.

COMPETITION

Plastics penetration of the paper and board markets has occurred particularly in packaging and wrapping applications. Low-density polyethylene film with over 1.5 million MT of consumption accounted for over 50% of thermoplastics materials in light packaging in Western Europe in 1972.[15] Table 1.5 lists the 1974 LD PE film consumption in heavy-duty sacks and the forecast for 1980 for the main West European countries and regions. According to this forecast made by Shell in 1976, the development of consumption of LD PE film in other major packaging applications in these countries will be as shown in Table 1.6.

With regard to the heavy-duty-sacks market there is a certain competition within the plastics industry. Woven polypropylene tape and woven high-density polyethylene-tape sacks, in some cases with 1-ply

Table 1.5 *Heavy-Duty Sack LD PE Film Consumption (in thousand MT)*

	1974	*1980*
Fed Rep Germany	85	103
UK	48	74
France	52	88
Benelux	48	68
Italy	35	45
Scandinavia	42	63
Switzerland	3	4
Total	313	445

15. Unpublished Study, Shell, 1974.

Table 1.6 *Development of LD PE Film Consumption (in thousand MT)*

	1974	*1980*
Refuse sacks	170	327
Carrier bags	195	286
Shrink and stretch film	225	504
Laminates	84	169
Miscellaneous packaging	681	958

kraft paper, have replaced jute sacks in many applications.[16]

Another plastics material for heavy-duty sacks is Valeron, a laminate consisting of two oriented high-molecular weight HD PE sheets laminated at an angle of 90°. This material has the properties of high impact, puncture, and tear resistance and of low vapor permeation. In thicknesses of 75 and 100 microns it competes against 5- and 6-ply kraft paper as well as against 250- to 300-micron LD PE film.

Developed and manufactured by Van Leer of Belgium and Van Leer Plastics (USA) Inc., Valeron is also supplied in bags. In Europe, Valeron is used for packaging carpets and textile rolls as well as for shipping synthetic fibers and polymers, thus competing with polypropylene-tape woven fabrics. In transporting heavy goods, however, Valeron is sometimes used as a liner in combination with this type of woven sack. St. Regis Bag Packaging Division (USA), reports that it supplies bags made from Valeron, and that among the users of these bags is American Gilsonite Co., which used multiwall bags for packaging of the asphaltite material before switching over toValeron. As a result of this switchover, water permeation has been eliminated, and breakage during ocean trips has been cut practically to zero. The Valeron bags also have the advantage of being about one-fifth the weight and occupying about one-fourth the space of multiwall sacks of the same capacity. Copper Pigment and Chemical Works Inc., a supplier of copper and nickel sulphates, has also replaced 6-ply paperbags with Valeron bags made by St. Regis. The company did this because the wet strength of Valeron bags is higher, and products packed in these bags can be stacked higher, thereby saving warehouse space.

Refuse sacks made from LD PE film have been competing with paper bags and, to some extent, with dustbins. It is significant that during the London dustman strike a few years ago, the local authorities issued LD PE bags, not kraft paper sacks, to the householders. In the case of

16. PP or HD PE extruded film is oriented and slit (or first slit and then oriented) into tape (flat yarn) which is woven into a sack material.

shopping bags, the wet strength of LD PE film is an important advantage over paper bags. It is possible that in the future HD PE bags will in some cases replace LD PE bags, since HD PE is stiffer and lower gauge material can be used.

Shrink and Stretch Films

Shrink film and, lately, stretch film have been successfully competing with paper and board, and at the same time new packaging systems combining trays made of board and shrink films have been developed.[17]

Though they vary from case to case depending on the efficiency of the previously used system, the actual material- and labor-cost savings achieved by firms that have replaced conventional board boxes (and kraft paper) are considerable. Supermarkets and cash and carry shops, which have been gaining a growing share of the retail trade, are an important contributing factor in the accelerating growth of shrink wrapping. Shrink wrapping is a more convenient form of packaging than board cases for these retailers. Shrink-wrapped parcels are easier to open, and the disposal of wrappings is more convenient and economical than the disposal of board, which is bulkier. Shrink-wrapped packs have a greater sales appeal at the wholesale and retail level, a point particularly stressed by cash and carry shops. It is significant that some leading papermakers have been using shrink film for packaging their products.

Shrink wrapping embraces a wide range of fields of application. Food products and nonfood consumer goods are shrink wrapped individually for display in shops. Some of these goods, including canned and bottled beverages, are also supplied and placed on shelves in shrink-wrapped multipacks. The dispatch of these goods has been at an increased rate on pallets, whereby the entire load and the pallet are shrink overwrapped. Many building materials are transported shrink-wrapped on pallets to the building sites where they can be left in the open, protected against rain and wind by the shrink wrap, which because of its transparency also allows an easy identification of the packed materials.

Some heavy loads are shrink wrapped without pallets (bundling). Depending on the kind and size of products, different types and thicknesses of shrink film are used. In Europe low-density polyethylene film is mainly used for shrink wrapping, but PVC shrink film competes with LD PE for transit packs in display applications where a high "sparkle" is required for sales-appeal reasons. In the USA and in Japan polypropylene is used for making shrink film to a larger extent than in Europe. Polyester film, PVdC film, and some other sophisticated films are used

17. Wolpert, *Shrink Film—A Growing Market.*

in the shrink-film market for special applications, including the meat and cheese trade.

Recently, stretch-film wrapping of pallet loads has been introduced. This system has the advantage that no oven or heat is required. There is also no danger of fusion between the stretch film and the inner PE shrink film used for wrapping of individual packs within the pallet load. However, the equipment involved is more expensive than shrink-film equipment, and the strength of the shrink wrapping is claimed to be stronger than that of the stretch film.

It is estimated that in Britain the consumption of LD PE shrink film together with a small amount of stretch film amounted to over 20,000 MT in 1974, and will reach an annual consumption of about 40,000 MT by 1980, with an increased share of LD PE stretch film (in the USA, stretch film has been finding a growing acceptance).

The UK consumption of LD PE shrink film is lower than the German consumption, which in 1973 had already reached more than 70,000 MT, the French consumption, which in 1974 reached more than 35,000 MT, and the Italian consumption, which in 1974 reached about 30,000 MT. And further consumption increases are expected in Western Europe, as previously mentioned.

Shrink-film hood SH-T on a palletized load. Courtesy Alkor-Oerlikon Plastic Co.

Shrink-film tunnel, built by Nütro Hirsch & Co., Nürnberg, Fed Rep Germany. Courtesy Nütro Hirsch & Co.

The consumption of PVC plasticized shrink and stretch films (mainly for food packaging) in Western Europe is estimated to have amounted to about 24,000 MT in 1974, with the UK being the biggest market with about 8,000 MT.

Foamed Plastics Sheet

Expanded plastics films and sheets possess a cellular structure that provides the finished article with greater stiffness for a given weight of polymer; they have thus been finding a growing market in packaging applications.

Expanded polystyrene sheet has secured a share of the market of trays, vending cups, egg boxes, etc., successfully competing with pulp-based materials. Thin expanded polystyrene sheet has been replacing paper and paperboard in table doilies, table mats, and plates.

Expanded polyolefin film used for wrapping Camay soap has the advantages of protecting the soap against moisture penetration and of retaining the soap's fragrance. This film, produced by H. P. Smith Paper Co. of Chicago, competes with coated paper and won the President's Award at the 1970 National Flexible Packaging Association Competition in the USA.

The six major markets of foamed polystyrene products in the USA are:

1. Egg cartons (8,000 million cartons were used in 1973).
2. Meat trays (it is estimated that 70% of all packaged meats in foodstores are sold in foamed polystyrene trays. Pulp and unfoamed polystyrene account for most of the balance).
3. Produce trays, including for packaging apples and pears.
4. Disposable foodware (plates, etc.).
5. Dunnage.
6. Liners for wet surface containers.

This $400-million-a-year industry competes with products of the pulp industry as well as with china plates in hospitals, cafeterias, etc.[18] (It is estimated that the cost of washing and drying a 9-inch china plate in a hospital amounts to about 7 cents, whereas a 9-inch plate made of foamed polystyrene, which can be used once and thrown away, costs between 3 and 5 cents.)

Gulf States Paper Corporation (USA), which has been active in papermaking for nearly 90 years, has diversified its activities into the plastics field. The company has been producing lunch trays for school cafeterias from expanded polystyrene sheet because they feel that the functional properties of these plastics trays are superior to those made of pulp.

Du Pont estimated that by 1978 the six major markets just mentioned will be consuming 600 million lb of extruded polystyrene foam annually in the USA.

Polystyrene Laminates

Thermoformed laminates of XPS (expanded PS) and TPS (toughened polystyrene) have been finding application in the market of institutional trays. Table 1.7 shows the properties of this type of XPS-TPS laminates.

Polystyrene foam sheet to which kraft paper or other type of paper is laminated on one or both sides offers interesting possibilities in packaging. In Europe Shell/Berstorff and LMP/Colombo have developed processes and equipment for the manufacture of these laminates.[19]

Products made by the LMP/Colombo process are called Laminil. The thickness and weight of these laminates can be determined in accordance with the requirements of the specific application. The weight and

---

18. *Du Pont Magazine* (International Edition), Geneva, Switzerland, May 1974.

19. For Shell/Bertstorff, see Wolpert, *Synthetic and Plastics Paper*, Vol. 2, pp. 243-46; for LMP/Colombo, see Vol. 3, pp. 520-23.

Table 1.7 *Properties of Expanded Polystyrene Sheet and Expanded Polystyrene Sheet Laminated with Carinex Toughened Polystyrene*

Material	Total thickness mm	TPS thickness mm	Weight kg/m^2	Tensile yield stress Mn/m^2	Elongation %	Tear res. N (g)	FWIS joules (gf m)	Fold endurance cycles	Load to yield kg	Flexural yield strength Mn/m^2	Flexural modulus Mn/m^2 (kgf/cm^2)	Peel strength kg (a)	Cell size MD μm	Cell size TD μm
1	1.06	–	0.11	3.07 (31.2)	5.15	1.99 (204)	0.07 (7.14)	58	0.09	2.43 (24.8)	1.3 (13.2)	–	270	150
2	1.45	0.21(b)	0.28	3.59 (36.6)	7.15	2.34 (238)	0.31(c) (31.6)	531	0.34(c)	4.67(c) (47.5)	3.0(c) (30.6)	0.47	–	–
							0.36(d) (36.7)		0.24(d)	3.37(d) (34.4)	2.7(d) ((27.6)			
1	1.11	–	0.12	3.28 (33.4)	6.30	2.35 (239)	0.06 (6.12)	38	0.14	2.92 (29.8)	1.4 (14.3)	–	210	150
2	1.44	0.20	0.30	3.68 (37.6)	7.90	3.19 (325)	0.37(c) (37.7)	976	0.37(c)	4.94(c) (50.4)	2.2(c) (22.4)	0.54	–	–
							0.42(d) (42.8)		0.19(d)	2.45(d) (25.0)	1.9(d) (19.4)			
1	1.21	–	0.12	3.28 (33.4)	6.15	1.96 (200)	0.10 (10.2)	44	0.13	2.56 (26.2)	1.36 (13.9)	–	190	140
							0.22(c) (22.4)		0.38(c)	4.17(c) (42.3)	1.98(c) (20.2)			
2	1.66	0.13	0.28	3.11 (31.8)	4.90	2.90 (296)	0.46(d) (46.9)	1584	0.19(d)	2.20(d) (22.4)	1.73(a) (17.6)	0.61	–	–

Note: 1 = expanded polystyrene, 2 = laminate, MD = machine direction, TD = transverse direction.

(a) Peel 25 mm wide strip crosshead speed 25 mm/min.
(b) There is a discrepancy between the thickness of basic sheet and laminate minus toughened polystyrene owing to post-expansion of the expanded polystyrene.
(c) Toughened polystyrene upper face.
(d) Toughened polystyrene under face.

Source: Shell Chemicals U.K. Ltd., "Laminating Expanded Polystyrene Sheet," *Styrocell Technical Manual*, Section 2.62, March 1973, p. 2.

A paper laminated polystyrene sheet, Laminil, made by Livorazione Materie Plastiche (LMP), Torino, Italy, and converted by Cartiere Burgo, Italy, is used here for packaging soft drinks. It boasts high lightness and shock absorption properties. Courtesy LMP.

cost of the laminate is claimed to compare favorably with corrugated cardboard. Boxes made from these laminates could offer significant advantages in transporting foodstuffs from developing countries and could provide an important new outlet for the polymer and paper industries, replacing certain conventional materials such as plywood for tea-chests.

These extruded expanded polystyrene sheet-paper laminates are also expected to achieve a market penetration in the production of boxes for export of fish, transport of fish from port to retailer, transport of vegetables (cucumber, lettuce, cauliflower) from sorting house to retailer, export of fruit (apples) and tomatoes, export of margarine, air transport of flowers, transport of frozen foods, packaging things like high-value deep-frozen cream gateaux, and replacement of fiber-board boxes in heavy shipping containers for canned goods.

It is stressed that expanded polystyrene sheet-paper laminates have the advantages of retaining their strength at high humidities, being of light weight (a laminate box will have the same strength as a corrugated box that weights 20% more), having twice the insulation capability of a corrugated board of similar quality, having high puncture resistance, and possessing excellent printability. In addition, these laminates show a considerable strength in both the machine direction and the transverse direction, whereas corrugated board is stronger in one direction only.[20]

VERSATILITY OF PLASTICS FILMS FOR PACKAGING

The wide range of packaging applications of plastics films is due to their versatility and the variety of methods available to process them into finished packages, including such processes as heat sealing, bonding, stitching, etc. Some films can be vacuum-formed into trays.

Plastics films provide on the whole a better protection of the packaged goods than paper. And consumer goods and foodstuffs packaged in transparent plastics film have a greater sales appeal than those packed in paper.

PP film has been successfully competing with cellophane film in applications such as breadwraps and merchandise wrappings (shirts, sweaters, and other textiles). Coated PP film is also used for wrapping of individual cigarette packs and over-wrapping of cartons (200 cigarettes in 10 packs). In the case of American-style cigarette soft packs, PP film provides dimensional stability; the packs do not crush so easily as they do with cellophane film. PP film is used as packaging material for snack foods, potato crisps, etc. In addition to single-ply PP film, which provides an efficient barrier against the ingress of moisture, PP film is also used extensively in laminates (e.g., PP film/PP film, PP film/cellophane film, PP film/PE film, the latter possessing high strength and puncture resistance). Table 1.8 shows comparative data for some transparent packaging films.

Chisso Corporation, Tokyo, which supplies PP resin Polypro and PP tubular and flat extrusion plant, lists in its brochure (*Chisso's Blown Film Manufacturing Process*) the following comparative data, which show that PP (specific gravity of about 0.90) has a lower specific gravity than the following materials:

Film	*Specific Gravity*
Cellophane	1.40–1.50
PVC soft	1.25–1.40
PVC hard	1.40–1.45
HD PE	0.93–0.96
LD PE	0.91–0.93

20. Shell Chemicals (UK) Ltd., Technical Leaflet T3351/4.73/750, "Laminating Expanded Polystyrene Sheet," March 1973.

Table 1.8 *Comparative Data on Some Transparent Packaging Films*

Property	*Units*		*15-micron*[1] *Propafilm O*	*25-micron cast polypropylene*	*25-micron low-density polyethylene*	*Cellulose films* 350 MS	*Cellulose films* 365 MXXT
Density	g/cm^3		0.91	0.90	0.92	not homogeneous	
Yield	m^2/kg		73.4	44.0	42.5	28.4	27.4
	in^2/lb		51,500	31,000	30,000	20,000	19,300
Tensile strength	kg/cm^2	MD	1,750-2,100	450	200	900-1,350	900-1,350
	lb/in^2		25,000-30,000	6,500	2,800	13,000-19,000	13,000-19,000
	kg/cm^2	TD	1,750-2,100	390	150	420-560	420-560
	lb/in^2		25,000-30,000	5,500	2,200	6,000-8,000	6,000-8,000
Elongation at break	% of original length	MD	50-85	750	300	20	20
		TD	50-85	1,000	650	60	60
Tear strength	g/0.001 in	MD	5 approx.	50	200	5 approx.	5 approx.
		TD	5 approx.	80	150	8-12	8-12
Modulus of elasticity	kg/cm^2 x 10^{-4}	MD	2.2-2.7	0.75-1.0	0.2-0.25	3.0-5.7	3.0-5.7
	lb/in^2 x 10^{-5}		3.2-3.8	1.1-1.4	0.3-0.35	4.4-8.3	4.4-8.3
	kg/cm^2 x 10^{-4}	TD	2.2-2.7	0.75-1.0	0.2-0.25	1.5-2.7	1.5-2.7
	lb/in^2 x 10^{-5}		3.2-3.8	1.1-1.4	0.3-0.35	2.2-3.9	2.2-3.9
Melting point	°C		168-171	168-171	107-112	–	–
	°F		335-340	335-340	225-234	–	–
Permeability to water vapor	g/m^2/24 hr at 90% rh at 38°C (100°F)		8-10	12	17	5-12 (uncreased)	1-8
Permeability to oxygen	cm^3/m^2/24 hr at 1 atm differential pressure		3,000-4,000	4,500 approx.	9,500 approx.	5-25	5-15

Source: ICI, Packaging Films Sales Department, brochure entitled "Get Tough."

Note: MD = machine direction, TD = transverse direction.

1. Oriented PP film.

A significant feature of the plastics industry is that various polymers compete in the same applications. Polyester film that is glossy and transparent competes with polypropylene film in various markets and finds application either alone or in laminates for packaging of foodstuffs, liquids, medical products, etc. Laminates of polyester film and polyethylene are used for boil-in-the-bag applications. New coated films and laminates, including those with nylons, ionomers, and PVdC, are finding new applications in packaging. At the same time, in some cases these packages use paper, thus providing more paper industry outlets.

ELECTRICAL INDUSTRY

In insulation applications, plastics films have secured a share of the market, replacing impregnated paper. In this field various types of plastics films compete with paper and with each other. For example, in the production of coils, plastics films, cellulose triacetate film, and cellulose butyrate-based cast film have all secured a share of the market.

In addition to plastics films, which are finding an increasing market in the electrical industry, spunbonded (see Chapter 3) and allied materials have been introduced for some applications within this industry. For instance, Du Pont's Reemay, a spunbonded polyester sheet, has an interesting potential in coated and laminated structures, including for electrical and filtration uses.

Nomex nylon by Du Pont is supplied in the form of paper, as well as in the form of filament yarn and staple fiber, and although it is not a spunbonded product, it has been added by Du Pont to its group of spunbonded materials. Nomex paper possesses a valuable combination of good electrical properties, a long useful life at temperatures above 220°C, and a very high mechanical strength. Electrical uses of this product include:

- ☐ Turn, layer, barrier, section, lead, and tap insulation utilized in transformers.
- ☐ Conductor, coil, slot, phase, wedge, and end insulation in motors and generators.
- ☐ Printed circuit backings, terminal and potentiometer strips, cable insulation, creped and pressure-sensitive tapes.
- ☐ Spiral and convolute-wound tubes, coil bobbins, V-rings, and a variety of other formed parts.

According to Du Pont, short, uncrimped staple can be combined with small fibrous binder particles of Nomex polymer (called fibrids) to produce sheet structures by using conventional papermaking techniques. After formation, these two constituents are permanently bonded together by high-temperature calendering without the use of additional binders, fillers, or sizes. By blending mica into Nomex paper, a special product has been developed which is claimed to be superior to mica splittings for some electrical uses.

Capacitors

Biaxially oriented polypropylene film has achieved a significant penetration of the capacitor market, replacing paper. Other plastics films, including biaxially oriented polyester film, polystyrene film, polycarbonate film, and polysulfone film have also secured a share of this market for some types of capacitors. Trials with high-density polyethylene film are reported to be in progress by some leading capacitor manufacturers.

The replacement of paper in the production of capacitors, which are now being made at an increasing rate either entirely from plastics films or from systems made of part paper and part plastics, has the following significant features:

1. In the case of capacitors, plastics films replace high-price types of tissue paper (approximately £1,500-2,000 MT).
2. The impetus for this development came from the electrical industry, since plastics film capacitors have superior functional properties and the size of the capacitor has been reduced.
3. Several European manufacturers of tissue paper for capacitors have installed equipment to produce plastics films for this application. In some cases they have secured a license from General Electric Co. of America, which has been very active in this field and holds a number of patents related to the production of films and capacitors.

In the high-voltage power-factor-correction field, capacitors utilizing part paper and part plastics (mainly polypropylene) are being used, and systems where solid dielectric consists entirely of plastics film are on the market. Whereas cellulosic paper reaches the maximum tolerable power loss because of thermal instability at stresses below 20 kV/mm, polypropylene film has been used successfully at nearly three times the stress.[21] Plastics films also have the advantage of high uniformity and have only a very few thin spots.

21. N. Parkman, Electrical Research Association, at the "Plastics and Paper" Conference (organized by the Plastics Institute), London, May 1973.

In the manufacture of DC capacitors, the conventional use of paper has been replaced to a large extent first by biaxially oriented polyester film and later, in some countries, by polycarbonate film, which replaced part of the polyester film market. In the AC-capacitor market, polycarbonate metallized film (partly in combination with paper) was introduced, resulting in smaller-size capacitors. But this film secured only a comparatively small share of the market. Later, polypropylene film, which is considerably lower in price, began to penetrate this market at a high rate.

Biaxially oriented PP film with a specific gravity of about 0.9 (which is lower than that of most other dielectric materials) provides a larger area of dielectric per weight unit. An 8-micron PP film weighs 7.2 g/m^2. Bulk availability and low price, coupled with high performance, give PP a high growth rate potential in this market.

In Western Europe it was estimated that PP film consumption for impregnated capacitors would increase from about 500 MT in 1972 to about 2,400 MT by 1977. It was also estimated that PP film would find an even bigger market in dry capacitors, where consumption was expected to increase from approximately 850 MT in 1972 to approximately 3,500 MT in 1977.

Polyester film consumption in Western Europe for capacitors is expected to have a growth rate of about 5% per year, using as a base the 1971 consumption of about 1,300 MT. About 90% of the polyester film is used for electronics capacitors and the remaining 10% for industrial capacitors.

According to General Electric Company (USA) polypropylene capacitor-grade films have the following market trends:

1. *Electronic film capacitors*. Single sheet plain and metallized polypropylene capacitor lines have been replacing many polystyrene, polyester, and polycarbonate designs in entertainment, communications, instrumentation, and computer segments; the high performance and low cost of the PP film have made possible many new successful product lines.
2. *Industrial film capacitors*. The use of PP film is growing rapidly in this area. Plain PP and PP/paper impregnated capacitors are forecast to replace dual paper designs in the lighting, motor run, high energy, industrial electronics and other industrial segments.
3. *Power capacitors*. It is estimated that over 85% of all power capacitors are constructed with PP film. The polypropylene Magvar design has reduced losses by 50%, increased dielectric strength by more than 200%, and has greatly assisted designers in obtaining more joules of energy per unit weight.

General Electric feels that continued research with polypropylene is providing a basis for further design improvements with the "all polypropylene" power capacitor design forecast, and they feel that the biggest new market will be an all-film-impregnated PP capacitor that will replace the multiple-sheet paper capacitor used in motor run, lighting, and electronic AC capacitors.

Manufacturers of specialized tissue paper for capacitors have reacted in two ways to the market penetration by plastics films. Some of them (e.g., Bolloré, Tervakoski) have installed production lines for making special types of plastics films for capacitors. Others feel that despite the plastics film penetration of this market, the overall demand will continue growing at a high rate, and the demand for paper will thus not diminish. New specialized types of paper have been developed, including those for the metallization of these ultra-high density papers that are highly glazed with a surface very similar to those of plastics films.

According to Julius Glatz GmbH (Fed Rep Germany), a leading manufacturer of tissue paper for capacitors, the 1971 consumption in Western Europe was estimated at about 11,000 MT. They pointed out that despite the increasing use of PP films for capacitors, there is still a very great demand for capacitor papers.

In order to keep their share of the market, many industries that came under pressure because of new synthetics have increased their research and development (R & D) activities. This increased R & D activity has benefited the industries in question as well as the industries using their products.[22]

CABLE INDUSTRY

Plastics have achieved a high penetration of the wire and cable industry's market, and it is estimated that in the USA about 100,000 tons of LD PE were used in cable applications in 1974.

Since crosslinked PE has a higher maximum conductor temperature than high-molecular-weight PE, the crosslinked PE (XLPE) has secured about 85% of the installation of the secondary-distribution power cables, and about 47% of the high-voltage power cables (above 600 volts). It is estimated that in 1975 the share of LD PE accounted for 82% (crosslinked LD PE—47%, high-molecular-weight LD PE—35%) of the total solid insulation materials for power-cable insulation (utility—above 600 volts, mining and industrial power cables). EPR (ethylene-propylene rubber) accounted for 15% of this total market. And the share

22. See Wolpert, *Synthetic and Plastics Paper*, Vol. 1, pp. 79-87; Vol. 3, pp. 469-98; Vol. 4, pp. 662-67.

of this market in which paper finds application has gradually declined from 10% in 1963 to 3% during the last few years.

STATIONERY AND DRAWING OFFICE APPLICATIONS

In stationery and drawing office applications, plastics films have begun a certain market penetration at the expense of paper and cellophane products. Oriented PP and oriented PS film have been finding application in window patching of envelopes. Carbon "paper" based on PP film of about 19-micron thickness (yield about 54 sq m/kg) has been competing against conventional carbon paper and also against the earlier introduced "paper" based on polyester film of about 12-micron thickness (yield about 59 sq m/kg). Although the price of both these plastics films is higher than that of conventional paper, this higher price is justified by their better performance and longer "working life." (According to Du Pont, carbon film made from polyester provides up to five times the service of conventional carbon paper.)

In Britain the annual consumption of PP film for this application is estimated at over 250 MT and that of polyester film at about 200 MT. PP film, even on an area basis, is cheaper than polyester film. The share in Britain of plastics-film-based carbons accounts for about one-third of all multi-use manufactured carbons and for about two-thirds of the exported carbons. It seems likely that this share will continue to rise and that many countries which now seem to be slow in accepting these plastics film products will eventually turn more to them at the expense of the conventional wax oil carbons.[23]

In Britain polyester films have achieved an important penetration into the drafting-materials market, and according to ICI Plastics Division, drafting material based on its Melinex polyester films have "removed many of the limitations imposed by paper and linen."[24] The main advantages of these films are dimensional stability, durability, long life (these films do not contain plasticizers which might migrate, and in normal use do not embrittle), a higher degree of translucency, and a high resistance to indentation.

In Fed Rep Germany it is estimated that about 1,200 MT of highly transparent heliographic paper and about 2,400 MT of tracing paper for technical drawing purposes are consumed annually. Some specialty papers for this application consist of an inner layer of polyester film, to

23. A. S. Hickman, Ozalid Co., Ltd., UK, at the "Plastics and Paper" Conference, London, May 1973.

24. "Melinex Films in the Drawing Office," Technical brochure, ICI Plastics Division, Welwyn, Garden City, England, September 1969.

which a highly transparent tracing paper is laminated on both sides (manufacturer: Gebr. Hoesch, Kreuzau), representing another case of cooperation between the two industries.

Data processing cards made of laminates (two layers of paper with a polyester film core) have been introduced in Fed Rep Germany and Holland. Other applications of such laminates and/or plastics films are jacquard and dobby cards. In some cases, the functional properties of these products justify their higher costs (dimensional stability is particularly important in areas where climatic conditions, such as dampness, have an adverse effect).

2. Definition of Different Types of Synthetic and Plastics Papers

The penetration of some traditional paper and board markets by plastics is not a new phenomenon. This trend has been observed in the growing packaging industry, in refuse-bag production, in certain fields of the electrical and electronic industry (cables, capacitors), in the wallpaper industry, etc. Until recently, however, these plastics products were never described as "paper" products. The new chapter of the competition began with some products of the plastics industry being introduced as "synthetic paper" and "plastics paper." There is still confusion about classification of some of these products, although it has been discussed at various international conferences, including the IUPAC-EUCEPA conference at Helsinki in June 1972, where the author of this book introduced the classifications discussed in this chapter.[1]

SYNTHETIC PAPER

Synthetic paper comprises a wide range of products that aim at specialty markets of the graphic industry; these products can be broadly subdivided into those based on plastics extruded film, on synthetic fibers, and on spunbonded materials.

Based on Plastics Extruded Film

Plastics extruded film (of various polymers) undergoes various types of post-extrusion treatment (clay coating and/or filling with titanium dioxide, silicon dioxide, calcium carbonate, etc.) in order to achieve

1. See *Man-made Polymers in Papermaking: Proceedings of the IUPAC-EUCEPA Symposium, Helsinki, June 1972* and Wolpert, *Synthetic and Plastics Paper,* Vol. 1.

"paper resemblance" (opacity, rigidity, creasing ability, and good printability). This process is sometimes called "paperizing" of plastics film. A number of patents have been granted for the surface treatment of plastics films by mechanical and chemical means, resulting in a better ink receptivity, opacity, etc. Some of these films (filled prior to extrusion) are biaxially oriented and do not require any surface coating. There are also multi-ply extruded films that are oriented, and the total production is done in one continuous process.

Among the latest developments are films extruded from a mixture of HD PE (high-density polyethylene) and other polymers. For some applications these films are coated, on one or both sides. When the films are not coated, they represent a borderline case between synthetic papers and plastics papers.

Based on Synthetic Fibers

Synthetic paper products based on synthetic fibers and special types of rayon, mainly in combination with conventional pulp, have been on the market since before the advent of plastics-film-based products. They are made on conventional papermaking machines, and in Europe they find application in the production of drivers' licenses, official documents, and geographical maps—markets in which they have been lately exposed to competition from plastics-film-based synthetic paper. In the USA some manufacturers have developed specialized latex-saturated papers, including those with synthetic fiber/wood fiber furnish with special latex systems.

Rayon/polypropylene tissue for tea bags is known in the tea trade as "synthetic" paper, whereas manila hemp tissue with a layer of heat-sealable fibers from a vinyl chloride/vinyl acetate copolymer is described as "natural" paper.[2] On the European continent, tea bags are also from a furnish of 70% hemp or chemical straw pulp plus 30% thermoplastic fiber (85% vinyl chloride and 15% vinyl acetate), the latter providing wet-strength reinforcement.

Based on Spunbonded Materials

Spunbonded materials aim mainly at textile markets, but some coated linear polyethylene spunbonded materials are finding an increasing market in labels, tags, book covers, wall coverings, envelopes, etc. These products belong to the overall group of synthetic paper.

2. I. G. Lewis, of Lyons Tetley, Ltd., in "Tea Bag Tissue," (Presentation at the "Plastics and Paper" Conference, London, May 1973).

SYNTHETIC PULP PAPER

Synthetic pulp is a new material for papermaking. Synthetic pulp, which is made by a continuous process from synthetic monomers (e.g., ethylene, propylene, etc.), and which consists of very fine fibrils, is processed to paper on a conventional papermaking machine. Paper can be made either from 100% synthetic pulp or from a furnish consisting of synthetic and conventional pulp, depending on the required properties of the end product. At present, five semicommercial pilot plants of about 3,000-10,000 MT/yr capacity each—applying this production method—are in operation. Other technologies for producing synthetic pulp are at the laboratory stage and are described in Chapter 4.

At present synthetic pulp is considered to be an additive to the furnish of conventional wood pulp for making paper, etc., and thus represents a further aspect of cooperation between the synthetic polymer and the paper industries.

PLASTICS PAPER

Plastics paper, or paper-like high-density polyethylene film, which on the whole does not require any post-extrusion treatment, has been competing with paper (and to some extent with low-density polyethylene film) in wrapping and bag-making applications. In contrast to other plastics films (e.g., low-density polyethylene [LD PE] film, which has been competing with some types of kraft paper, and various types of high-density polyethylene [HD PE] film, which have always been described as plastics films), this type of HD PE film, first introduced in South Africa about seven years ago, has sometimes been called *paper-like film, papier plastique,* and *plaper.* In Germany the term *HM Folien* (i.e., high-molecular-weight film) is often used, even when the film is made from medium-molecular-weight HD PE. The term *paper-like* refers to the "paper handle" of this film, its stiffness, good foldability, and mat surface, but this film also has properties that paper does not possess: wet strength, tear resistance, weldability, gas impermeability, etc.

Although plastics belong to the overall group of "synthetics," it is appropriate to make a distinction between "synthetic' and "plastics" papers, because of the differences in their production methods (plastics paper does not require any post-extrusion treatment and at present does not aim at printing- and writing-paper markets).

3. Synthetic Papers

Synthetic papers aim at the printing- and writing-paper markets, with an emphasis on specialty types of these papers. The magnitude of this market and its growth outlook can be seen from the data in Tables 3.1–3.3. Printing and writing papers (newsprint excluded) account for over 22% of the total production of the paper and board industry. (Table 3.1 shows the 1974 production of printing and writing paper, newsprint excluded, broken down by OECD regions.) It is generally expected that the consumption of printing and writing paper will continue to grow, with its share of total paper and board consumption remaining about 22%.

The OECD forecast that the total production capacity of the paper and board industry in the OECD regions will increase from 130.1 million MT in 1973 to 158.1 million MT in 1978.[1] And according to the FAO Report of March 1972, which introduced for the first time the consideration of a profitability constraint on the production outlook, the production-constrained outlook for consumption of printing and writing paper (newsprint excluded) was estimated as is shown in Table 3.2. The FAO Report of March 1974 estimated the growth of the rated printing- and writing-paper production capacity as shown in Table 3.3.

According to an early 1976 forecast by English China Clays (UK), a major supplier of filler and coating materials to the paper industry, the apparent consumption of printing and writing paper (newsprint excluded) in Western Europe will have an annual average growth rate of 6% between 1976 and 1980. This forecast takes into account the 1974-75 recession, and it estimates that in 1977 and 1978 an over-average growth rate will be registered.

Synthetic papers were introduced with the aim of penetrating the markets of printing papers and writing papers. Although the various types of synthetic papers differ in the raw materials from which they are produced and in the applied production methods, they have properties in

1. OECD, *The Pulp and Paper Industry, 1974-1975* (Paris: 1976).

Table 3.1 *1974 Production of Printing and Writing Paper (Newsprint Excluded)*

	OECD production (million MT)
Western Europe	12.5
Canada	1.1
USA	12.0
Japan	2.9
Australia and New Zealand	–
Total	28.6

Source: Compiled from OECD, *The Pulp and Paper Industry 1974-1975*, Paris, 1976.

common that are superior (to a varying extent) to those of conventional pulp-based papers, particularly with regard to tear resistance, water repellency, chemicals resistance, dimensional stability, and folding endurance.

The impulse to produce a material to replace conventional paper was stimulated by the threatened shortage of pulp. Japan, a country poor in natural resources, with a strongly increasing paper and board consumption, was in the forefront of the development of synthetic paper based on plastics film. Japan's expanding pulp and paper industry was (and still is)

Table 3.2 *Outlook for Consumption of Printing and Writing Paper (in million MT)*

	1969	*1975*	*1980*	*1985*
Western Europe	7.91	10.80	14.00	17.00
North America	10.56	13.05	16.55	20.05
Japan	1.85	3.10	4.52	6.07
Oceania and South Africa	0.39	0.54	0.73	0.97
USSR	1.23	2.20	3.10	4.50
Eastern Europe	0.86	1.28	1.85	2.40
China	0.85	1.50	2.05	2.65
Africa	0.13	0.14	0.26	0.42
Asia and the Far East	0.82	1.35	2.65	2.80
Latin America	0.76	1.20	1.45	1.95
Middle East	0.12	0.22	0.40	0.60
World	25.48	35.38	46.96	59.41

Source: FAO, *Outlook for Pulp and Paper Consumption, Production and Trade*, March 1972.

Note: 1969 mean of 1968, 1969, and 1970 (1970 preliminary figures only).

Table 3.3 *Growth of Printing- and Writing-Paper Capacity (in million MT)*

	North America	*Western Europe*	*Japan*	*World total*
1973	13.1	12.8	3.1	35.6
1975	13.7	14.1	3.6	38.6
1977	14.3	15.3	4.0	41.6
1978	14.4	15.7	4.0	42.6

Source: FAO, *World Pulp and Paper Capacities – 1973-1978, FAO Survey, 1974*, March 1974.

faced with the very serious problem of raw materials supply. The Japanese felt that from the point of view of their national economy it would be better to import additional quantities of crude oil (which was in abundant supply in the 1960s) than to be dependent on deliveries of raw materials for paper manufacture. Japan's main suppliers of these raw materials have been the North American countries, where prominent suppliers have issued warnings of possible shortages. A recent stockbrokers' report pointed out that the U.S. forest products industry has been moving south, so as to free-up West Coast lands to serve Japan (at astronomical prices).[2]

In May, 1968, Japan's Resources Commission Recommendation stressed the following points:

> The demand for paper is expected to exceed 16 million tons by 1978, and about 1,100 million US $ of foreign currency would be needed if these demands are met by importing pulp.
>
> Foreign currency to the value of about 500 million US $ will be needed should Japan rely on imported wood.
>
> Already by 1967 imports of timber were valued at about 930 million US $.
>
> If the degree of dependence on domestic resources is raised, it is feared that this will become a serious threat to "territorial integrity", through devastation of the forests.[3]

The recommendations also referred to the fact that the cost of "petroleum as basic raw material for the production of synthetic paper is about one quarter to one half" the cost of the pulp and timber necessary for the production of conventional paper, based on the same area. And they stressed that synthetic paper production would provide a large tonnage market for Japan's petrochemical industry, which must be expanded in order to increase its international competitive strength.

2. Mitchell, Hutchins Inc., *Status Report*, New York, January 18, 1974.

3. See Wolpert, *Synthetic and Plastics Paper*, Vol. 1, p. 142.

In 1972 the Japanese Government Agency of Science and Technology formed the Technology Assessment (TA) Committee on Synthetic Paper within the Natural Resources Council, and in October 1975, the technology assessment report was issued. This report covers the 1973–85 period, and refers to film-type synthetic papers, synthetic pulp papers, and highly pigmented papers. Questions regarding raw materials and their availability, pollution, etc., are discussed.[4]

The committee determined that the quantity of oil required in the production of synthetic paper was about twice that required for woodpulp paper, but that the difference decreased if one took into account the energy used for woodcutting, delivery, sea transport, etc.

One important factor discussed was that woodpulp paper consumes far more water and emits much more sewage than synthetic paper. It was therefore felt that with the possibility of technological improvements in the production of synthetic paper, this paper might prove to be more advantageous than woodpulp paper, taking into account Japan's scanty natural resources.

The Committee felt that the advanced stage of production of synthetic paper and of the utilization of plastics scrap will be reached by about 1980, when the output of synthetic paper will reach 100,000 MT/yr. At that time the technological assessment should be made again on the basis of gained experience.

It was also stressed that since the Japanese Government has not yet announced the estimated volume of the long-term oil supply (after the ''oil shock'', i.e. oil embargo and price increases at the end of 1973), this investigation should be updated as soon as the Government announces the long-term estimates.

Japan's production of paper and paperboard has increased since 1970, when Japan became the second biggest paper and board manufacturing country, and the third biggest woodpulp producing country, in the world, registering, however, due to the recession, a decrease in paper and paperboard production in 1974—as the following table shows.

Year	*Million MT*
1970	13.0
1971	12.9
1972	13.6
1973	16.0
1974	15.6

4. Some details of this report are described by Dr. Shozo Imoto, Director, Toppan Printing Co., Tokyo, and Member of the Natural Resources Council, at the TAPPI International Synthetic Papers & Pulps Symposium, Washington, D.C., March 31-April 2, 1976.

Table 3.4 *Japanese Imports and Exports of Paper and Board (in MT)*

	Imports	*Exports*	*Balance*
1970	100,000	508,000	408,000
1972	143,000	506,200	363,200
1974	435,400	682,300	246,900

Source: Compiled from OECD, *The Pulp and Paper Industry 1974-75*, Paris, 1976.

This increased production has resulted in increased large quantities of pulpwood and woodpulp imports by Japan. Woodpulp imports increased from about 730,000 MT in 1970 to 1,230,000 MT in 1974, a nearly 70% increase. (In addition to these pulp imports for papermaking, in 1974 Japan imported 278,000 MT of dissolving pulps.)[5] These increases of woodpulp imports were much higher than Japan's increases of woodpulp exports (exports of 12,000 MT in 1970 increased to nearly 50,000 MT/yr in 1971-73 period, and reached 220,000 MT in 1974). It is also significant that Japan's imports of paper and board have increased at a higher rate than Japan's exports. The resulting lower balance of the country's exports over their imports of these particular products is shown in Table 3.4.

According to some estimates the fiscal-year production of paper and paperboard dropped from 14.4 million MT during 1974 (ended March 31, 1975)—representing a production decrease of 13.2% from the 1973 output—to 14.1 million MT during 1975, and was to show a recovery to 14.7 million MT in 1976.[6] At the same time, it is expected that the production capacity of Japan's paper and board industry will increase from 18.6 million MT in 1974 to about 20 million MT by 1977-78.

The paper and board industry is concerned about its dependence on raw materials due to limits in the country's resources and increasingly strict environmental regulations. The industry's dependence on pulp imports is bound to increase.[7] The industry is therefore endeavoring to secure a regular supply of pulp by participation in pulp-making projects in Brazil with the proviso that half of the produced pulp will be shipped to Japan, as well as by participation in some projects in New Zealand, Indonesia, and the Soviet Union to secure a regular supply of pulp and of pulp wood and to lessen its dependence on North American supplies that

5. OECD, *The Pulp and Paper Industry:* 1972-1973, 1973-1974, and 1974-1975 (Paris: 1974, 1975, 1976).

6. Fuji Bank Bulletin, Tokyo, May 1976.

7. OECD, *The Pulp and Paper Industry 1974-1975* (Paris: 1976).

are considered "unreliable" at times of brisk pulp and paper production in North America. In fact, a certain diversification of supplies has recently been taking place, and imports of paper woodpulp from European OECD countries reached nearly 140,000 MT in 1974, accounting for about 11% of Japan's total imports. However, in boom periods Western Europe can be hardly considered a potential large-scale supplier of woodpulp to Japan.

During 1970-71 the recession in Japan's paper industry resulted in very low prices of conventional paper; in 1972 the prices continued to be low (in most cases under the 1970 level) and did not reflect the actual production costs. The recession that started in the second half of 1974 and continued throughout 1975 was more severe than the 1970-71 recession.

COMPETITION AND MARKETS

Unlike plastics papers and other plastics products, which have been penetrating the packaging markets because of their functional advantages and price competitiveness, synthetic papers aiming at the printing- and writing-paper markets were more expensive than conventional paper (being offered at up to 3.5 times the price of conventional paper). The reasons for this price difference included:

1. The very low prices of conventional paper (the price situation was aggravated during the recessions).
2. The comparatively high production costs of synthetic papers due to the small capacities of the installed plants.

These price differences have been too high for a penetration by synthetic paper of the larger-tonnage markets of printing and writing papers. The use of synthetic papers has been confined exclusively to speciality applications of small tonnages, where their functional advantages justify their higher prices. Most synthetic papers are products of the first generation, and some of their manufacturers have been carrying out research work to develop synthetic papers of the second generation, with the aim of broadening the market of these products. The annual increase of conventional printing- and writing-papers consumption in the OECD region estimated at about 5% during the 1977-80 period would mean an annual increase of over 1.4 million MT.

In order to penetrate the market of printing and writing paper, or at least some sectors of this overall market, synthetic paper must satisfy certain requirements that are already met by conventional paper. It must do this, however, without having the same properties as conventional paper. In fact, the advantage of synthetic paper is that it has different, often superior, properties. Here, price competitiveness is of great importance—that is, whether the client is prepared to pay the higher price for a better-quality product. Better properties *as such* are meaningless; they must be related to the required functional properties of the end product.

It is uneconomical to use materials that last longer than the products they serve. For example, some maps have to be reprinted as they get obsolete. The use of certain types of synthetic paper for these maps would mean that the maps would still be in very good physical condition at the time they had to be discarded. But there are also cases when maps, made from some type of conventional paper, have deteriorated, and have to be discarded, because of the low-grade properties of the material used, long before they need to be reprinted. These two examples show that from the economical point of view the properties of the material from which the end product is made and the required properties of the end product must have a proper correlation.

Paper for many publications does not need to be grease- and water-resistant or repellent. However, in the case, for instance, of industrial manuals used in factories, it is a great advantage if drops of oil can be wiped off without damaging the publication, as can be done with synthetic paper. Guides, maps, posters, and other publications for outdoor use have the advantage of being water- and weather-resistant when they are made from synthetic paper.

SYNTHETIC PAPERS BASED ON PLASTICS FILMS

Paperization or Achieving a Paper-like Surface

Plastics films have on the whole a transparent quality; in order to give them a paper-like appearance and paper-like characteristics, pigments or auxiliary chemical additives (e.g., titanium dioxide) can be added to the polymer in the film-making process. The nonabsorbent, very smooth surface of a plastics film results in poor ink receptivity. Therefore, a sharp image cannot be reproduced quickly by ordinary writing or printing unless the film has been treated for these applications by special "internal" methods or surface treatments. Internal methods include foaming, leaching, stretching, and compounding (filling). Surface treatments include various types of coating and embedding of particles, forming a "swollen layer" on the surface, surface foaming, roughening or abrasing, and embossing.

In the 1960s, Nippon Kakoh of Tokyo, a partner in Japan Synthetic Paper Company, developed a method that is claimed to be superior to the previously developed methods for producing paper-like articles by surface treatment of polymer film.[8] This technology was licensed to Crown Zellerbach. While the company's earlier method involved forming a swollen layer on the surface of the material with a solvent or swelling agent, the new method provides means whereby the bonding-surface bonding strength of the opaque roughened layer formed is further improved without detracting from the printing and writing qualities of the layer. At the same time, variations in gloss, pattern, and color can be induced on the surface layer.

In this method the treatment of the plastics film usually takes place by continuous operation, in three or more stages:

1. Treating the film with a swelling agent.
2. Treating the film with liquids that are lower in swelling power than that of the swelling agent, successively in decreasing order of swelling power.
3. Treating the film with a coagulating agent.

After the treatment with the coagulating agent, and while the film is still in a semicoagulated state, watermark patterns can be printed on the film. This method results in an opaque roughened layer on the surface of the film, with a high level of printability.

In the production of synthetic paper the use of fillers (e.g., Kaolin and calcium carbonate) results in a lower raw material cost, since the price of fillers is considerably lower than that of polymers. In addition, high loadings of these fillers substantially reduce the amount of pigments (e.g., titanium dioxide) required to make the films as white as high-grade coated traditional paper.

The fillers in some synthetic papers also have the function of being the nucleus for cavitation (required for ink receptivity in printing), as well as of making the plastics film stiffer. However, at the same time, they reduce the tensile strength, tear resistance, and impact strength of the finished material.

Many synthetic papers, it is felt, have too low a stiffness; this has also been one of the problems in finding suitable applications for paper made exclusively from synthetic pulp. It is significant that at present synthetic pulp is used only in combination with conventional pulp (see Chapter 4) and that some manufacturers of synthetic pulp have abandoned their

8. British Patent Specifications 1,175,047, published December 23, 1969.

immediate efforts to find markets for the application of paper made from synthetic pulp only.

Microporosity of Synthetic Paper

Among technologies to achieve microporosity of synthetic paper is one developed by Freeport Kaolin Company (USA). According to Freeport Kaolin, PE and PP films containing the company's OX-1 surface-modified Kaolin, when biaxially oriented, develop a microporosity to yield very white, highly opaque film or sheets (e.g., a 0.7 mil biaxially oriented polypropylene sheet had 90 opacity, was very white, and showed good print qualities).

Freeport Kaolin has developed a process of modifying the surface of the clay and by this method producing an organophilic pigment (Kaolin clays are hydrophilic materials by nature and are readily wet by water). The hydrophobic nature of the surface-modified Kaolin clay, OX-1, eliminates the "wick effect," by which moisture can penetrate organic systems containing hydrophilic pigments. The organophilic nature of the fine-particle-size OX-1 permits very high OX-1 pigment loadings in organic monomeric and polymeric systems, with little increase in viscosity. Dense, nonporous plastics and fluid high-solids-solvent coatings are possible with the use of OX-1, which derives from an extremely fine particle-size aluminum silicate. Films and fibers having a high opacity and a high degree of whiteness can be produced through an unusual dewetting effect that takes place during the orientation of polymers containing OX-1.[9]

Freeport Kaolin has been stressing that it has no intention to enter the field of synthetic paper production but is interested in selling its clay to film and paper manufacturers. At the present time the company is evaluating the market and selling small quantities to film and paper manufacturers. The range of application of OX-1 is very wide and has a potential in the production of synthetic paper, film, coatings, laminates, nonwovens, and disposables, as well as in the production of synthetic pulp.

The report *Synpulp Projects and Technology* by Dr. William H. Hoge refers to the Freeport Kaolin U.S. Patent 3,697,474 of October 10, 1972, "Organo-Clay-Polymer Compositions," and says that by the proper stretching of a film containing 20% filler (density 2.4 g/cm^3) and 80% polypropylene (density 0.9), it is possible to obtain a final product of 0.78 g/cm^3 density.[10]

9. See Wolpert, *Synthetic and Plastics Paper;* Vol. 2, pp. 276-78 and Vol. 4, p. 682.

10. Published by Paperfilm Associates, Flemington, New Jersey, USA.

Potential Markets

According to Freeport Kaolin, modified Kaolin (of the OX-type) can be used in olefin polymers as articles or components of articles intended for use in contact with food. This should generate additional interest in OX types of Kaolin and in films made of polypropylene plus this type of Kaolin.

Opaque films and fibers made from a compound of polypropylene and modified Kaolin are comparatively inexpensive materials and provide potential applications in the paper and packaging industries. The use of such films in laminates with substrates of lower brightness could result in the production of decorative coatings; these films could also be used as functional coatings to produce lightweight sheets with paper-like properties. Fibers made from such films offer a low-ash-content filler for improving opacity and brightness of paper products.

Freeport Kaolin has produced highly opaque, white plastics film and fibers by mixing the organophilic Kaolin and thermoplastic polyolefins and then extruding the compound.[11] The organophilic Kaolin component results in a dense, void-free extrudent. The subsequent orientation of the extruded sheet at a temperature somewhat below the softening point of the polymer produces a multiplicity of submicrometer voids both within and at the surface of the film.

A 0.9 mil (about 21.5 micron) biaxially oriented PP film, 7x draw in each direction, was produced from a 30-mil (about 720 micron) sheet containing 20% by weight of the organophilic Kaolin. This film had an opacity of 91.8%, a Tappi brightness of 77%, and a specific gravity of 0.78.

Thin films, 0.5 to 1.0 mils, with opacities in the 85 to 90 range, have been produced at specific gravities of approximately 0.5, whereas unfilled, oriented transparent PP film has a specific gravity of about 0.9.

The Use of White Opaque Films

White opaque film can be used as a coating for paper or paperboard. Its low weight (17.8 g/m^2) and high opacity and brightness make this film highly suitable as a decorative coating to cover low-brightness substrates. Transparent polyolefin films are printed by rotogravure processes after surface oxidation of the film. The printability of white opaque film is also improved by surface oxidation, and it is stressed that the

11. In the joint paper by Horton H. Morris and Paul I. Prescott, Freeport Kaolin Company, "White opaque plastic film and fiber for papermaking use," (Presentation at the American Chemical Society Conference at Atlantic City, New Jersey, on September 9, 1974), Paul Prescott described the latest developments in applications of the company's organophilic Kaolin-type OX.

Table 3.5 *Coated and Uncoated Opaque Polyolefin Films*

	Reference no.				
	1	*2*	*3*	*4*	*5*
Coating composition					
Pigment	Ultrafine clay, 70 parts	Ultrafine clay, 75 parts	OX-1, 70 parts	Ultrafine clay, 70 parts	Ultrafine clay, 75 parts
Adhesive	Starch, 30 parts (Penford 280)	Low viscosity protein, 25 parts	Latex, 30 parts (Airfax 400)	Starch, 30 parts (Penford 280)	Low viscosity protein, 25 parts
Type film	Polypropylene, containing 20% OX-1	Polypropylene, containing 20% OX-1	Polypropylene, containing 20% OX-1	Polypropylene, containing 20% OX-1	Polyethylene, containing 10% OX-1
Uncoated film data					
Basis weight (lb/3,300 ft^2)	9.4	9.0	8.0	10.3	10.7
Thickness (mil, approx)	0.6	0.6	0.6	0.9	1.1
Opacity, TAPPI (%)	83.4	80.9	82.1	89.9	89.8
Brightness, TAPPI (%)	69.1	68.5	69.0	68.3	88.1
Gloss, B & L (%)	93	93	88	95	15
*Coated film data**					
Total coat weight, coated 2 sides (lb/ream)	9.1	4.8	7.8	9.7	4.0
Total coated sheet thickness (mil)	1.0	0.8	2.4	1.3	1.5
Opacity, TAPPI (%)	84.9	82.8	96.5	90.4	90.3
Brightness, TAPPI (%)	76.0	77.9	79.2	78.0	88.1
Gloss (%)	51	41	19	44	30

Source: Freeport Kaolin Co., Georgia, USA.

* Film subjected to corona discharge before being coated.

letterpress process can be also used in the case of opaque films.

In addition to the possibility of using the white opaque film as "coatings" to be laminated to other substrates, it is also possible to use these lightweight films as "coating substrates." The availability of very lightweight, smooth, strong substrates having good opacity and brightness provides a number of potential advantages. The disadvantages of such films (e.g., limpness and need for special treatment and/or special inks) can be overcome by coating the films.

Table 3.5 lists the properties obtained when opaque films were coated using conventional coating pigments with several types of additives. It was necessary to oxidize the films in order to coat them with the water-based coatings. The opaque film coated with the clay-starch coating was strong enough to enable the use of heat-set letterpress inks in printing. Coating the plastics film improved the stiffness and "rattle" of the sheet and appears to offer an approach to producing plastics films with paper-like properties.

Fibers for Paper Production

Monoaxial orientation of the filled PP film results in the film being easily fibrillated, and it can then be chopped to the desired length to obtain opaque fibers. These fibers with wetting agents can be used with cellulosic fibers in aqueous slurries. The addition at low-weight percentages of the OX-1 organophilic Kaolin results in higher opacity and brightness of the finished sheets, which can be also heat-bonded. U.S. patents for preparations of the pigment, the fibers, and the opaque film have been granted to Freeport Kaolin Company.[12]

Freeport reports that the Kaolin pigment prepared in this way is organo-dispersible and yields free-flowing slurries at high solids in hydrocarbon systems. The dispersibility of the treated pigments in hydrocarbons and polyolefins is due to the reduction of the surface energy of the Kaolins, and it is possible to modify pigments to be dispersible in most organic systems.

The *Synpulp* report mentioned earlier also refers to other developments in this field. The Honshu Paper Company-Lion Oil composition, Calp, is a mixture of about 65% anydrous calcium sulphate and 35% polyethylene. Polymer Corp. Ltd., Canada, has developed a polypropylene $CaCO_3$ film with density of 1.3 g/cm^3 which can obtain a density of 0.8 after a fivefold stretching. (This is possible because the pigment particles appear to act as nuclei for micro pores which are formed in the sheet.)

12. Patents No. 3,697,474; 3,697,475; and 3,660,134.

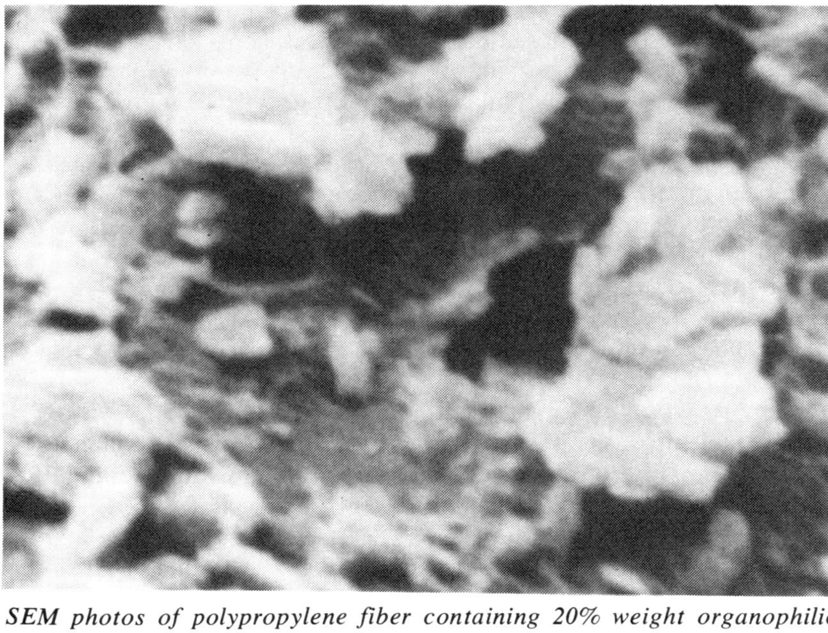

SEM photos of polypropylene fiber containing 20% weight organophilic Kaolin Ox-1: top, unoriented, at 23400x; bottom, oriented, at 22800x. The fibers have been oxidized to expose the pigment particles. Courtesy Freeport Kaolin Company.

MAIN TYPES OF SYNTHETIC PAPERS BASED ON PLASTICS FILMS

□ Filled, modified HD PE film based on synthetic papers.
  Polyart: manufactured by BXL-Bakelite Xylonite Ltd., England.
  Miraynart: manufactured by Mitsubishi Rayon Co., Ltd., Japan, under license from BXL (Polyart synthetic paper).
  Zexan: joint development of Mitsui Petrochemical-Sanwa Vinyl, Japan.
  Arathene: developed by UCB, Belgium.

□ Clay-coated polystyrene film based on synthetic paper.
  Q-Kote: developed by Japan Synthetic Paper Co., Japan.
  Phönix artplast: manufactured by Papierfabrik Scheufelen, Fed Rep Germany.

□ Solvent-treated polystyrene film based synthetic paper.
  Q-Per: developed by Japan Synthetic Paper Co., Japan.
  Spiax: developed by Sumitomo Chemical Co., Japan.

□ Solvent-treated PVC film based synthetic paper.
  Development by Neste Oy, Finland.

□ Water- and solvent-treated polystyrene film based synthetic paper.
  Printel S: developed by Sekisui Chemical Co., Japan.

□ Water- and solvent-treated PVC film based synthetic paper.
  Printel V: developed by Sekisui Chemical Co., Japan.

□ Multi-layered PP laminate based synthetic paper.
  Yupo FP: manufactured by Oji-Yuka Synthetic Paper Co. Ltd., Japan.

□ Filled-polyallomer film based synthetic paper.
  Scott Plastic Paper: developed by Scott Graphics Inc., USA.

The production of several types of synthetic papers referred to in this list has recently been terminated, because they did not secure a large enough market to be a profitable proposition, and also because of the 1974-75 recession. Manufacturers like Sekisui, however, feel that the work done on these synthetic papers enables them to continue their R&D activities in this field with the aim of developing synthetic papers of a second generation. Therefore, a description of these products (in addition to the historical interest of this industry's developments) has been included. Among products which have achieved a certain market penetration and are at present marketed are Polyart, Yupo FP, Phönix-artplast. This chapter also describes the developments of synthetic papers based on mineral-filled plastics films, HD PE/polystyrene compound films, and polypropylene films.

Filled, Modified HD PE Film - Synthetic Paper

Among the functional advantages of the filled, modified HD PE film-papers compared with conventional (cellulose-based) coated printing paper are:

- ☐ A very high (according to some claims up to 100%) ink holdout.
- ☐ No lint particles.
- ☐ No grain direction.
- ☐ Grease and water resistance.
- ☐ Water-proofness.
- ☐ Higher abrasion resistance.
- ☐ High air and gas impermeability.

Polyart

Polyart, developed and manufactured by BXL (Bakelite Xylonite Limited), England, is a film made from a modified high-density polyethylene that has been biaxially oriented. The product is not surface coated, but some additives are used, and the standard surface is semimatte on both sides. Polyart has a specific gravity of 1.14, and the standard types have the following weights and approximate thicknesses:

90 g/m^2 of about 3 mil thickness
115 g/m^2 of about 4 mil thickness
200 g/m^2 of about 7 mil thickness
(1 mil = 0.001 inch or about 25.4 micron)

Compared with conventional paper, Polyart has the advantage of impermeability to air and moisture; its water-vapor permeability is 14.4 g/m^2/24 hr/0.001 in (38°C, 90% RH, BS 3177). The mechanical properties of Polyart are those of a tough, continuous plastics film. Tearing is difficult since it involves disruption of the film itself rather than the separation of fibers that occurs with paper. Unlike paper, Polyart is unaffected by water (either as vapor or liquid). Differences between properties in the machine and transverse directions of manufacture commonly occur with papers and films. Such differences are very small with Polyart. Table 3.6 compares typical mechanical properties of Polyart with those of conventional art paper.

Commercial production of Polyart started in 1968, and it is sold in the UK through paper converters and merchants. In some European countries paper manufacturers and paper merchants market this product. In Austria, the paper company Neusiedler has been marketing Polyart. In Fed Rep Germany, Polyart is marketed by IGEPA of Reinbeck, under

Table 3.6 *Comparison of Mechanical Properties*

Property	*Units*	*Polyart*		*Art paper*	
		Wet	*Dry*	*Wet*	*Dry*
Yield[1]	lb/in^2	8,000	8,000	–	–
Breaking strength[1]	lb/in^2	9,500	9,500	160	5,000
Elongation at break[1]	5 in/min, %	140	125	4	4
Modulus[2]	lb/in thick	2.1×10^5	2.1×10^5	10^2	3.8×10^5
Initial trouser tear strength[2]	lb/in thick	500	500	NIL	50-100
Flex resistance[3]	No. of 180° folds without cracking	>50,000	>50,000	–	750-1,500

1. Test Method BS 1133.
2. Test Method BXL test.
3. Test Method BS 4419.

Source: Wolpert, *Synthetic and Plastics Paper*, Vol. 1, 1971, p. 92 (based on information supplied by BXL).

the trade mark PolyPrint; Siegwerk Farbenfabrik of Siegburg has been supplying ink for offset and letterpress printing on PolyPrint.

Leading UK ink manufacturers have formulated inks suitable for Polyart, and it is reported that, in the 115-g/m² grade, light transmission is 12% (test method: ASTM D1003-61) and printing opacity is 92%.

Among specialty applications of Polyart is its use on the new microrecord retrieval system produced by Micro and Cinebooks Ltd. of Jersey. Using a binocular lens assembly with double-image printing, a strong linear reduction has been obtained.

A nonexclusive manufacturing license has been granted to Mitsubishi Rayon Company of Tokyo to manufacture Polyart, which is marketed in Japan under the trade name Miraynart.

Polyart has been used for printing of various pocket guides, books, and manuals, particularly for outdoor use, as well as for making wall maps, survey and geological maps, outdoor posters, etc. The reference list of Polyart deliveries includes a growing number of customers using this synthetic paper of 200 g/m² for wall charts and also for the production of mathematical games. Polyart of 115 g/m² has been used for stickers and other applications.

Polyart is being marketed in Britain by Wiggins Teape Paper Co. (manufacturers of high-quality and specialty papers); BXL has similar agreements in some other countries with paper industry firms that have a thorough knowledge of the market. Furthermore, Samuel Jones & Co. Ltd., a subsidiary of Wiggins Teape, has been converting Polyart into

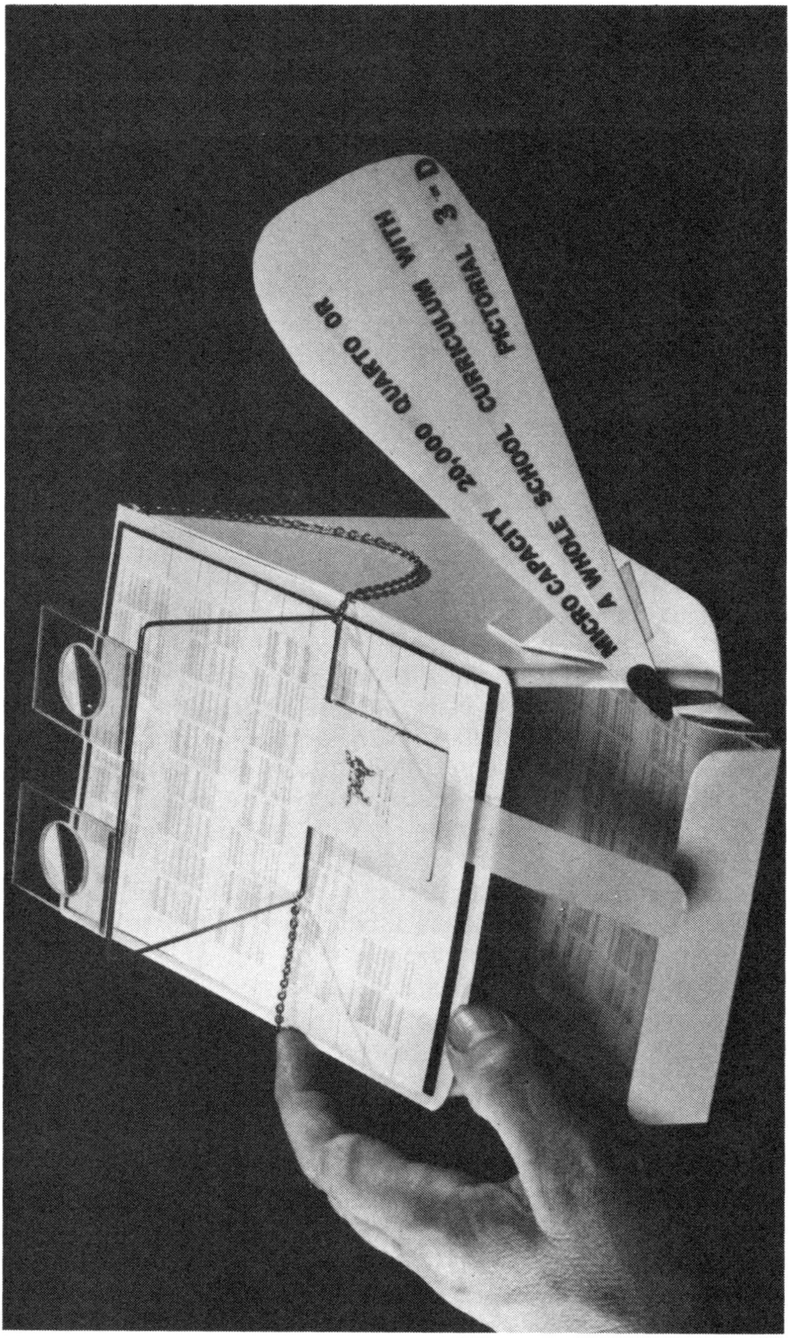

Micro record retrieval system of Micro and Cinebooks Ltd. of Jersey uses Polyart synthetic paper of BXL, Bakelite Xylonite Ltd., UK. Courtesy BXL.

Polyad by coating it with a specially formulated self-adhesive, and laminating it with a release-coated backing sheet. Polyad has been developed mainly for outdoor stickers and other applications where water, weather, and grease resistance are required. It competes in some applications against PVC, and Samuel Jones & Co. feels that Polyad, which can be printed by litho or letterpress and other conventional processes, looks more attractive.

Applications of Polyart outside the graphic industry include:

- ☐ Making bearing blocks on which the polyester magnetic tape runs in cassettes (manufactured in Holland). Translucent Polyart of 97 g/m^2 of 100-micron thickness has been chosen for this material's good frictional properties.
- ☐ File index systems.
- ☐ Tape measures from 115-g/m^2 Polyart.

Dr. T. Watson and L. Leese, BXL, in a joint paper "Development of Synthetic Papers Based on Biaxially Oriented Film" described Polyart as a good representative of first-generation synthetic papers.[13] They felt that the development of new second-generation materials would broaden the penetration of the printing and graphic industry's markets by synthetic papers, because of the lower potential cost and the improved printing properties of the new products. Among the economic features of plastics film-based synthetic papers was the possibility of manufacturing them in production units of 3,000-4,000 tons/yr. These production units have the advantage of operational flexibility.

The present high cost of and greater care required in printing synthetic papers of first generation have confined their use to specialty applications. Some of these products have been accepted by printers and end-users. Owing to increasing demand, including from the European Continent (about 60-70% of the Polyart output is exported), the production facilities of Polyart were expanded in 1975.

Zexan

Zexan, developed jointly by Mitsui Petrochemical Industries and Sanwa Vinyl Ind. Co. Ltd. of Tokyo, belongs to the category of filled HD PE synthetic paper. It has been developed from Mitsui's Hi-zex HD PE polymer. The base film is T-die extruded, and fillers including titanium dioxide are used. In addition, a special surface treatment is applied.

13. Presentation at TAPPI International Synthetic Papers & Pulps Symposium, Washington, D.C., March 31-April 2, 1976.

Zexan can be offset printed, and the wide range of potential applications includes adhesives tapes, first-aid bandages, posters, stickers, and "paper" bags. Zexan can be heat-sealed. Table 3.7 gives the physical properties of Zexan.

Arathene

Arathene, developed by UCB and manufactured by UCB-Sidac Division, Ghent (Belgium), is a synthetic paper based on tubular (blown) extruded film made from a blend of thermoplastics (mainly HD PE) and

Table 3.7 *Physical Properties of Zexan*

		Grade		
Test item	*Unit*	*(1)*	*(2)*	
Thickness	microns	104	70	
Weight	g/m^2	107.8	69.0	
Density	g/cc	1.04	0.99	
Hunter whiteness	%	89.6	89.8	
Opacity	%	92.0	86.3	
Gloss 75°		4.1	6.4	
Yield stress				
MD	kg	3.2	1.9	
TD	kg	3.0	1.7	
Tensile strength at break				
MD	kg	2.4	1.5	
TD	kg	2.7	1.4	
Elongation at break				
MD	%	60	50	
TD	%	49	30	
Tear strength				
MD	g	80	40	
TD	g	170	100	
Film impact strength	kg	480	290	
Folding endurance				
MD	times	7,400	>10,000	JIS P-8115
TD	times	4,270	7,000	
Stiffness				
MD	g	13.0	8.6	JIS P-8143
TD	g	13.9	8.6	
Smoothness (Bekk)				
front	sec	636	836	JIS P-8119
back	sec	1,417	1,244	

Source: Mitsui Petrochemicals, Tokyo, Japan.

Note: MD = machine direction; TD = transverse direction.

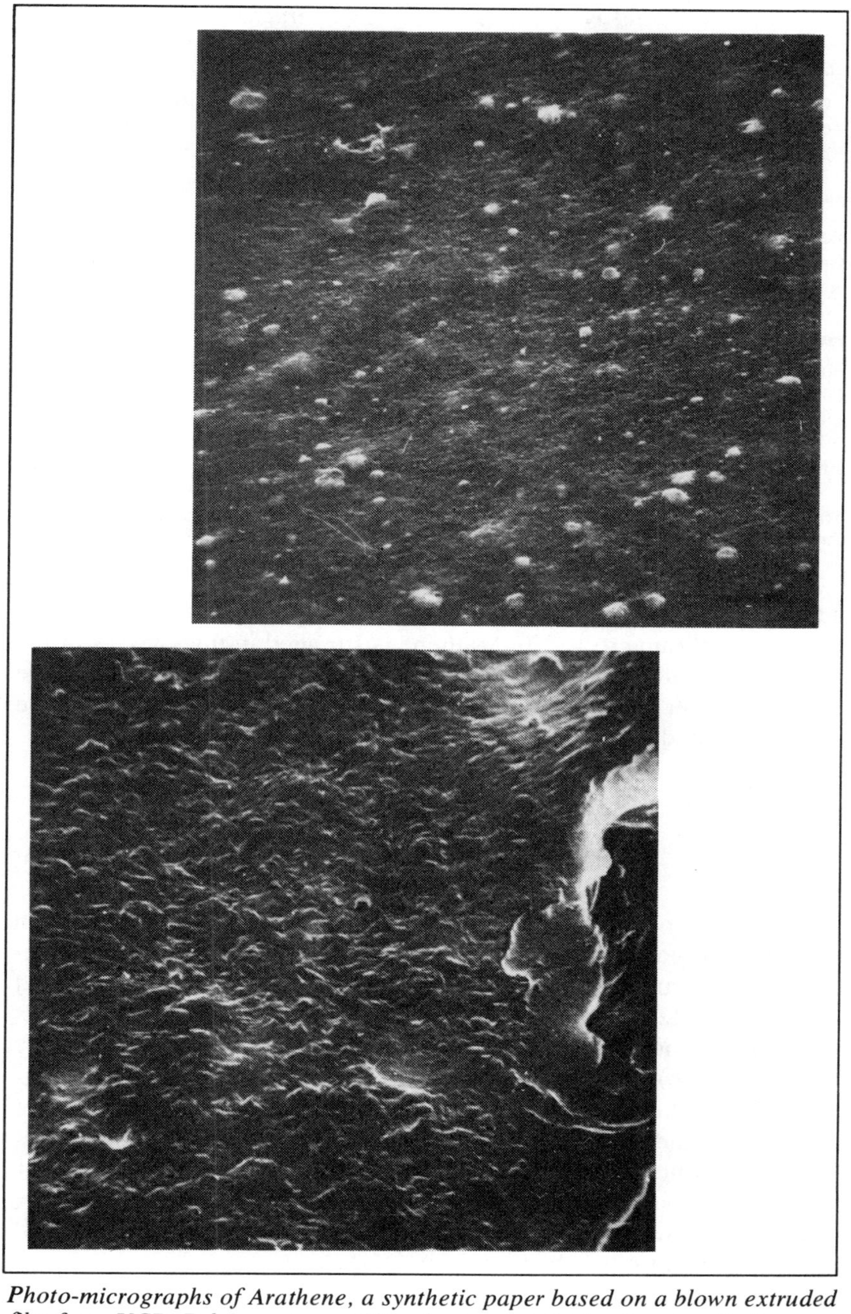

Photo-micrographs of Arathene, a synthetic paper based on a blown extruded film from UCB, Belgium: top, at 525x; bottom, at 4500x. Courtesy UCB.

Table 3.8 *Mechanical and Chemical Properties of Arathene*

Mechanical – for 180 micron film	*MD*	*TD*
Tensile strength	200 kg/cm^2	100 kg/cm^2
Tearing strength	50 g/sheet	150 g/sheet
Elasticity modulus	8,500 kg/cm^2	8,000 kg/cm^2
Chemical		
Water vapor transmission	3 g/24 hr · m^2 (38°C, 90% RH)	
Gas transmission (air)	30 cm^3/24 hr · m^2 (1 atm, 20°C, 40% RH)	
Resistance to greases and oils	excellent	
Permanence		
Maximum use temperature	+70°C	
Dimensional change at high relative humidity	none	
UV resistance	good	
Flammability characteristics	slow burning	

Source: UCB-Sidac Division, Ghent, Belgium (quoted in Wolpert, *Synthetic and Plastics Paper*, Vol. 2, 1972, p. 265).

additives (including titanium dioxide) using specially designed equipment.[14] According to UCB, Arathene is supplied in thicknesses of 80, 120, and 180 microns; since its specific gravity is 1.0, the weights of the synthetic paper are 80, 120, and 180 g/m^2 respectively. Arathene has the mechanical and chemical properties shown in Table 3.8.

Clay-coated Polystyrene Film Based Synthetic Paper

Q-Kote

Q-Kote, a pigment-coated specially treated plastics film has been developed by Japan Synthetic Paper Co. Ltd. Though polystyrene film has mainly been used in this application, other plastic films can be used as the substrate. The surface has an over-coating pigment layer and is very smooth and glossy. Japan Synthetic Paper Co. Ltd. was formed by Nihon (Nippon) Kako Seishi, holder of the patent on plastics-film treatment to achieve a paper-like surface; Asahi Dow, a manufacturer of polystyrene and other polymers; Jujo Seishi, a leading paper manufacturer; Dai Nihon Ink, a company that has evaluated special inks for printing on plastics films; and Mitsui Bussan, a company that has been marketing these synthetic papers. Table 3.9 shows the properties of Q-Kote with various substrate films.

14. Belgian Patent No. 752,917, July 3, 1970.

Table 3.9 *Properties of Q Kote with Various Substrata Films*

	Unit		*Biaxial oriented high-impact PS*	*Non-oriented PVC*	*Oriented polyethylene tere-phthalate*	*Non-oriented poly-carbonate*	*Non-oriented PE*	*Non-oriented PP*	*Foamed PE*
Basic weight	g/m^2		121	176	166	148	133	105	170
Thickness	microns		111	122	115	112	118	95	358
Density	g/cm^3		1.09	1.44	1.44	1.32	1.13	1.10	0.48
Brightness	%		87.5	86.8	87.0	86.5	86.7	86.0	88.0
Opacity	%		96.8	97.0	96.9	96.7	96.8	96.8	97.0
Smoothness	mmHg		22	20	18	21	33	35	23
Gloss	%		23	26	27	25	20	19	18
Tensile strength	kg	MD/TD	5.6/6.7	7.7/8.2	22/22	8.0/7.8	3.6/3.5	6.5/3.5	2.1/1.7
Elongation at rupture	%	MD/TD	35/24	13/10	40/40	48/15	100/58	90/40	65/48
Tear strength	g	MD/TD	59/30	92/80	136/128	96/93	83/120	35/21	52/42
Folding endurance	times	MD/TD	2,000/2,000	2,000/2,000	2,000/2,000	2,000/2,000	2,000/2,000	2,000/2,000	400/1,000
Burst strength	kg/cm^2		3.6	4.4	11	4.1	2.5	9.2	1.8

Source: Japan Synthetic Paper Co. Ltd., Tokyo, Japan.

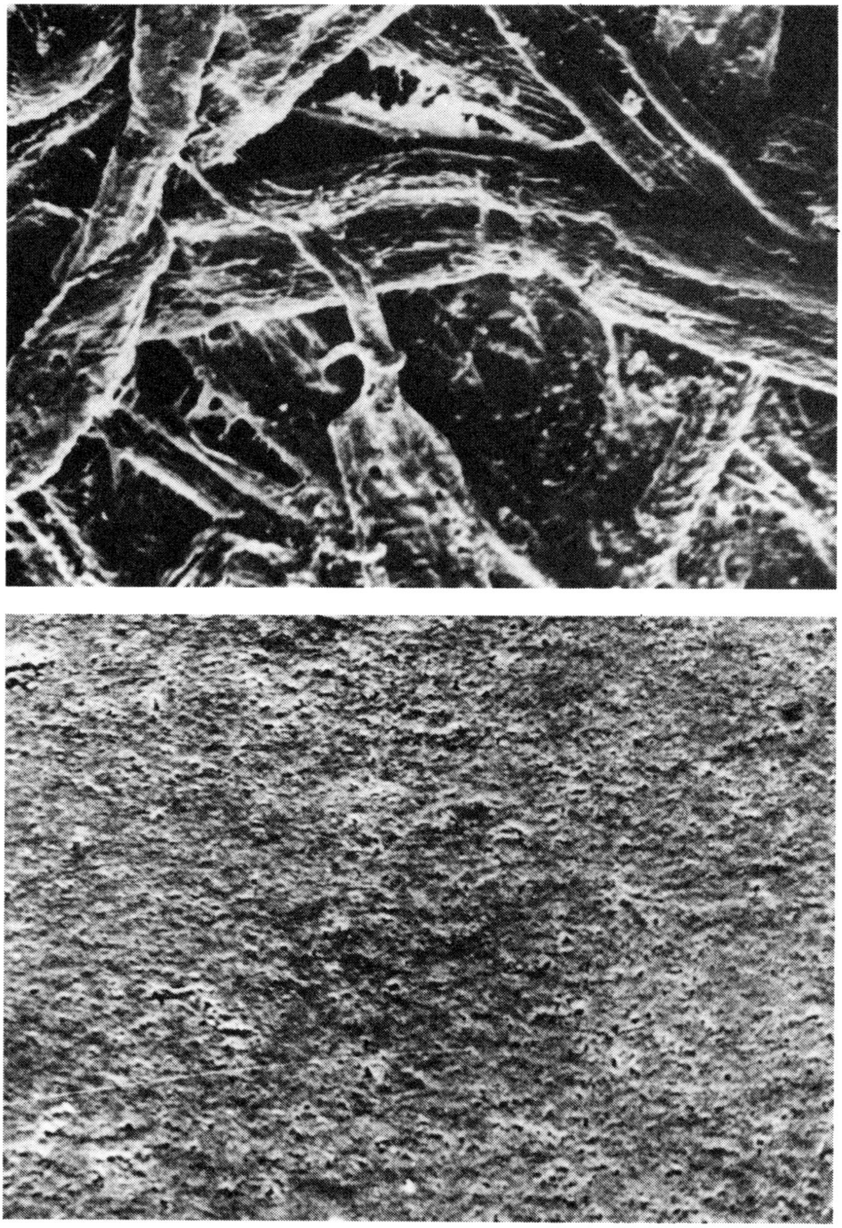

Phönix artplast is the first West German synthetic paper based on polystyrene. Scanning photo micrographs at 500x indicate excellent coverage possible by pigment coating: top, uncoated sheet; bottom, coated sheet. Courtesy Papierfabrik Scheufelen.

Q-Kote and Q-Per (described later) aimed at markets of high-quality commercial printing, and it was claimed that multicolor high-grade printing on these synthetic papers was superior to that on conventional coated paper. Q-Kote and Q-Per were among Japan's first commercially available synthetic papers with good printability characteristics.

Leading European and U.S. manufacturers had concluded arrangements with Japan Synthetic Paper Co. for evaluation of these products in the markets of their respective countries, with the proviso of possibly manufacturing these products.[15] However, the difference in price between synthetic papers and coated art paper has proved to be too high for securing a market of a size which would justify the establishment of manufacturing facilities. The import of these products from Japan has been at such a low level that most paper manufacturers who had signed agreements with Japan Synthetic Paper Co. have relinquished these arrangements. Similar experience has occurred with imports of Printel synthetic paper made by Sekisui.

Phönix Artplast

In May 1972, Papierfabrik Scheufelen (Fed Rep Germany), introduced the first German synthetic paper based on PS film—under the trade mark Phönix artplast. This synthetic paper, which is a biaxially oriented film and is double-pigment-coated on both sides, is directed toward the specialty products market, for use in such items as posters, labels, maps, and book covers. It has been supplied in 90 g/m^2, 130 g/m^2, and 180 g/m^2 weights, and according to Scheufelen, conventional inks and printing methods can be used for printing on it. The company's development of Phönix artplast followed the R&D work done by BASF.

BASF Development Work

BASF (Fed Rep Germany) has carried out extensive research work on the utilization of PS for making synthetic paper. It seems that the company's intention was to find an additional market for its PS polymers and not to enter into the production of synthetic paper. BASF's 1971 Annual Report said that a "promising use of polystyrene includes synthetic paper. Compared to conventional paper, the dimensional stability

15. Crown Zellerbach was marketing Q-Kote under the trade name Celestra-Kote and Q-Per under the trade name Celestra (see *Man-made Polymers in Papermaking: Proceedings of the IUPAC-EUCEPA Symposium, Helsinki, June 1972* [Helsinki: Finnish Pulp and Paper Research Institute, 1973] pp. 328-29). Sixteen pages and the cover of the April 1971 issue of the American monthly journal *Today's Packager* were printed on Celestra-Kote synthetic paper produced by Crown Zellerbach under license from Japan Synthetic Paper Co. (see Wolpert, *Synthetic and Plastics Paper*, Vol. 3, p. 424). Celestra-Kote was also used for making labels.

and wet strength of plastics offer distinct advantages."

Synthetic paper, based on BASF biaxially expanded 475 K polystyrene film, pigment-coated with Acronal S 320 D as binder, was used for cover pages of various issues of "Kunststoff Schnelldienst", a special publication distributed at K-71 Exhibition (Düsseldorf, September 1971). BASF stressed in its special Technical Information Sheet (Ti-SP.Obd, issued August 1971) that synthetic paper, because of its dimensional stability and wet strength, is a valuable addition to traditional pulp-based paper, and that recently developed production methods have resulted in plastics films with mechanical and printability properties similar to those of conventional paper. BASF felt that polystyrene, because of its easy processibility and its then low price, occupied an important position in the field of synthetic papers.

In addition to applications in the graphic industry's market, the polystyrene paper of transparent type aims at the tracing-paper market; in this application it has the great advantage of dimensional stability, a highly important property.

Manufacturing Methods

The production of the polystyrene-film paper from granules consists of three stages:

1. Extrusion.
2. Orientation of the film.
3. Treatment of the film for improvement of its printability.

Extrusion can be carried out by the conventional flat or tube method. Orientation of the extruded film improves the mechanical properties (stiffness, tear resistance, rupture resistance). The stretch ratio of 2:1 to 4:1 in both directions has proved to be easily obtainable. For the treatment of the film in order to improve its printability, the Corona treatment is recommended. Pigment coating of the film is similar to that of conventional paper; however, special attention should be paid to the selection of binders. BASF recommends its Acronal 290 D (possibly mixed with Acronal 500 D). The pigmentation should achieve higher opacity and better printability characteristics. It is recommended to use Kaolin together with titanium dioxide.

Pigment Coating

The joint paper by I. Reinbold and H. Lunk on "Pigment Coating to Improve the Printability of Plastics Films" described various methods of manufacturing coated plastics films based on polystyrene, and the authors felt that not every type of plastics film is suitable for pigment coating.[16]

Table 3.10 shows the mechanical properties of two types of conventional paper for art printing, and of various types of polystyrene films without internal pigmentation. The table shows that despite structural differences, the mechanical properties of paper and of extruded films are on the whole comparable. On the one hand, conventional papers are superior to plastics films in that the latter have a more excessive elongation at break and a lower tear propagation. On the other hand, plastics films are superior to paper in that they have a greater uniformity of properties in the machine and cross directions, and they also have a higher folding endurance and higher stiffness.

The table also clearly shows the importance of biaxial orientation of the films. Orientation makes general-purpose polystyrene suitable for printing. In the case of high-impact polystyrene types, orientation results in a savings of approximately 30% in material, which means that films can be approximately 30% thinner.

Referring to polystyrene films with internal pigmentation, the authors remarked that PS films are not normally very opaque. In applications for the graphic industry, it is possible to incorporate pigments (e.g., titanium dioxide) when the films are manufactured, instead of subsequent pigment coating, in order to achieve the necessary opacity. However, the inclusion of titanium dioxide affects the mechanical properties of the film.

Since conventional inks (used in printing on paper) do not adhere well to untreated polystyrene films and are very slow to dry, it is necessary to improve the printability of such films either by roughening the surface of the film (mechanically or chemically) or by applying a pigment/binder coating composition to the film (in the same way as coating of paper is done). According to the authors, only the second method provides a particularly high-grade quality of print, such as can only be attained on papers that are surface-coated with a pigment coating.

In contrast to the Sumitomo method of producing PS-film-based synthetic paper by solvent-treatment of the film, the authors stressed the point that for surface treatment pigment coating is less expensive than solvent treatment. They also pointed out that the mechanical properties of a polystyrene film are changed with the application of pigment coating, and that stiffness properties and tear-propagation properties are improved by the pigment coating. They felt that pigment coating is necessary in cases of high-quality printing.

16. Presented by I. Reinbold, BASF, at the IUPAC-EUCEPA Symposium, (see *Man-made Polymers in Papermaking: Proceedings of the IUPAC-EUCEPA Symposium, Helsinki, June 1972*, pp. 365-93).

Table 3.10 *Mechanical Properties of Conventional Papers and of Unoriented and Oriented Polystyrene Films without Internal Pigmentation*

	Degree of orientation		Substance (g/m^2)	Tear length (km)		Elongation at break (%)		Initial tear strength (kg)		Tear propagation (cmg/cm)		Folding endurance (times)		Stiffness (P)	
	MD	TD		MD	TD	MD	TD	MD	TD	MD	TD	MD	TD	MD	TD
Coating base paper (wood containing)			66	4.3	2	1.5	2.4	0.5	0.8	69	92	17	5	69	31
Coating base paper (wood free)			63	6.1	3	1.7	5.4	0.8	0.9	95	102	98	40	60	32
Polystyrene (general purpose)			118	5	4.4	2.6	4.8	0.6	0.8	not measurable		0	0	220	180
Polystyrene (medium impact)			119	3.7	2.4	28	26	1.1	1.4	35	80	609	47	84	83
Polystyrene (high impact)			101	2.4	1.8	57	57	1.5	1.6	51	73	3,360	1,600	66	46
Polystyrene GP (oriented)	2.5	3.2	105	7	6.5	5.5	5.4	2	2.4	59	71	225	105	440	360
Polystyrene (oriented medium impact)	3.5	3.4	104	4.6	4.3	16	24	1.9	2.1	83	99	4,050	2,620	220	180
Polystyrene (oriented high impact)	3.3	3.5	79	4.8	4.8	94	106	2.1	2.3	77	73	2,500	2,000	150	130

Source: Data from I. Reinbold and H. Lunk, "Pigment Coating to Improve the Printability of Plastics Films," *Man-Made Polymers in Papermaking: Proceedings of the IUPAC-EUCEPA Symposium* (Helsinki: Finnish Pulp and Paper Research Institute, 1973), pp. 365-93.

During the discussion following their presentation, it was mentioned that this kind of synthetic paper is based on a water-impermeable plastics film, but that the clay coating could be removed or affected by water. Also, the coating of plastics films eliminates the possibility of reusing the scrap unless the coating is removed first. Reinbold replied that this paper was developed for applications which required pigment coating.

Solvent-treated Polystyrene Film Based Synthetic Paper

Q Per

Q Per, developed by Nihon Kako Seishi and manufactured by Japan Synthetic Paper Company Ltd., is made from plastics film, the surface of which has been "paperized" (solvent treated). This paper is suitable for various applications. Q Per has been offered in various types:

QG: the surface of this type has a good gloss similar to that of high-grade coated paper.
QT: translucent paper, which finds application as tracing paper, photocopy paper, luminescent display paper, etc.
QM: matte paper, which has a high opacity and can be "transparentized" by applying heat or pressure to the desired areas of the surface.

Q Per has been developed on the basis of polystyrene film, but other plastics films can be applied in the production of this material.

Spiax

Spiax, the synthetic paper developed by Sumitomo Chemical Company Ltd., Japan, was supplied in two series:

FS series: white and opaque.
HS series: half-transparent, mainly for tracing paper.

The production process of both series of Spiax consists of high-impact PS being blended with fillers and then extruded into a sheet which is biaxially oriented into a film, the surface of which is solvent treated so that the properties of the product can be influenced by the degree of blending with fillers, the stretching ratio, and the solvent resin.

FS Spiax can be offset printed with conventional printing machines and ink, and it is claimed that antipicking in offset printing is "excellent" because of its monolayer (noncoated type) structure.

Table 3.11 *Properties of Spiax*

Testing item	*Testing specification*	*Unit*	*FS-110*	*FS-250*	*HS-70*	*Art paper*
Thickness	JIS P8118	micron	110	250	70	120
Basis weight	JIS P8124	g/m^2	130	295	77	125
Brightness	JIS P8123	%	90	90	80	90
Opacity	JIS P8138	%	92	98	45	97
Gloss	JIS Z8741 (60°-60°)	%	12	9	6	23
Smoothness (Bekk)	JIS P8119	sec	1,500-2,500	1,000-2,000	200	600
Tensile strength	JIS P8113	kg/mm^2	40/40	3.1/3.3	4.0/3.8	5.0/3.0
Elongation	JIS P8113	%	10-20/10-20	20-30/10-20	10-20/10-20	4/8
Tear strength	JIS P8116	%	14/15	>25/>25	11/11	20/18
Clark's critical length	JIS P8143	cm	45/40	70/90	20/20	60/65
Folding endurance	JIS P8115	time	>5,000	>5,000	>5,000	120/60
Coefficiency of heat contraction	80°C, 60 min in the gear oven	%	0.0	0.0	0.0	–
	130°C, 30 min in the gear oven	%	60-70/60-70	50-60/60-70	60/60	–
Weatherability	Weather-O-Meter		>200 hr	>200 hr	<100 hr, change in color	<100 hr, change in color

Source: Sumitomo Chemical Company Ltd., Osaka, Japan.

HS Spiax has been developed for tracing paper, copying and copyable paper, various types of data processing cards material for coating, etc. Table 3.11 lists the properties of Spiax and compares them with those of "art paper."

Sumitomo's invention to produce a translucent synthetic paper based on a styrene polymer, useful for drafting paper, original drawings, and copying, has been patented.[17] This product, made from a composition of 100 parts by weight of a styrene polymer and from 3 to 40 parts by weight of a white inorganic filler, is prepared by mixing with a solvent having a shrinkage initiation time of 40 to 2,000 seconds. The film has a heat shrinkage of from 2% to 90% in at least one axial direction.

The translucent paper of this invention has an opacity of 10% to 65% (measured in accordance with Japanese Industrial Standard P-18128), and it is claimed that this synthetic paper has good "writability"; that is, indicia can be clearly written on the paper in pencil and can be erased without damaging the surface of the paper, and the paper will satisfactorily accept writing in ink or Indian ink without blotting. The patent specification stresses the point that styrene polymer films incorporating the inorganic filler but not treated with the mixed solvent have inadequate writability characteristics and usually do not have an opacity higher than 10%. The translucent synthetic paper produced by this process also has good mechanical strength and can be used for preparing a photosensitive paper for copying.

The conventional pulp/paper for this application is produced by prolonged beating of chemical pulp, which is then subjected to the normal papermaking process, followed by a specific treatment to impart the desired transparency. The patent specification claims that translucent papers prepared from pulp cannot satisfactorily be used for repeated copying (which is their principle application), because they have poor mechanical strength and dimensional stability under moist conditions, and their low moisture resistance also precludes their storage for prolonged periods of time. This is a good example of certain advantages of the functional properties of synthetic paper over those of conventional paper.

Solvent-treated PVC film

Neste Oy, Finland, has developed a synthetic paper based on PVC film which is made by calendering or by solution casting. The film is composed of PVC and other ingredients, such as talc, Kaolin and possibly with cellulosic ethers or esters, polystyrene or ABS-resins. The

17. British Patent Specification 1,324,063 "Synthetic Paper," published July 18, 1973.

Table 3.12 *Comparison of Neste Oy Paper and Conventional Coated Art Paper*

Properties	*Unit*	*Neste Oy paper*	*Coated art paper*
Thickness	micron	110	70-90
Whiteness	%	87	82-88
Opacity	%	97	93-97
Breaking length			
MD	km	4	5-8
TD	km	2.2	2-3
Stiffness			
MD	g	8	12-24
TD	g	9	5-10
Folding endurance			
MD	times	1,500	30-150
TD	times	900	15-50

Source: Neste Oy, Finland.

"paperization" of the base film consists of immersion of the film into a solution of a polar and nonpolar organic solvent, of partial drying, of monoaxial or biaxial stretching, and of the final foaming at an elevated temperature.

According to some unconfirmed reports the film is heavily filled with calcium carbonate, which considerably reduces the production cost of this synthetic paper. Table 3.12 compares the properties of its synthetic paper with those of conventional coated art paper of the same weights (90-100 g/m^2).

Water and Solvent-treated Plastics Film Based Synthetic Papers

In 1970 Sekisui Chemical Company Ltd., Japan, introduced synthetic paper Printel, which is based on various types of plastics films: Printel S on polystyrene film, Printel V on PVC film, Printel E on polyethylene film.

Printel S

Printel S, which was first produced in 1970, found some applications in outdoor posters, wallpaper, tags, labels for cans, pressed sheets, etc. According to 'General survey of synthetic paper films' by Dr. R. W. Wright and H. H. Walbaum (International Paper Company), Printel was made by using water and solvent-treated film, and Printel S was "manufactured by water elution of a biaxially oriented polystyrene film containing poly(ethylene oxide), calcium carbonate and titanium dioxide pigments. . . . Solvent/non-solvent and foaming treatments disclosed in Sekisui patents [U.S. Patent 3,551,538—1971] may also have been applied in the manufacture of the product. To produce a porous surface, the film is impregnated with a foaming agent prior or after the elution with water. On the surface of the material, some particulate matter can be distinguished resembling spherulites which probably originate from the crystalline poly(ethylene oxide).''[18]

Printel V

Printel V consists of ''poly(vinyl chloride) modified with poly(ethylene oxide). The amorphous polymer is biaxially oriented for greater stiffness. The polyethylene oxide present in the extruded sheet is again removed by a water leaching process and . . . . appears to be partially redeposited on the surface. However, the presence of titanium dioxide or even clay is quite possible.''[19]

Sekisui told the author of this book (early 1976) that the company has suspended the marketing of its general purpose synthetic papers (Printel), and has been concentrating its efforts in the field of high-value products using synthetic paper technology and utilizing the experience gained in the production of the synthetic papers of the first generation.

Multi-layer Synthetic Paper

Yupo FP

Yupo FP, a three-ply PP synthetic paper manufactured by Oji-Yuka, is often described as a ''second generation'' product.[20] In May 1973, Oji-Yuka announced the conclusion of an agreement with Kimberly-Clark Corporation (USA) granting the latter exclusive sales rights in North America and Central America for Oji-Yuka ''polypropylene durable materials.'' This agreement included an option for a joint venture or licensing agreement for manufacturing this particular synthetic paper in the USA.

Oji-Yuka is a company founded in 1970 on a 50:50 basis by Mitsubishi Petrochemical Company and Oji Paper Company. It is an example of

18. Dr. R.W. Wright and H.H. Walbaum, International Paper Company, (see *Man-made Polymers in Papermaking: Proceedings of the IUPAC-EUCEPA Symposium, Helsinki, June 1972,* pp. 327-50).

19. Ibid.

20. Wolpert, *Synthetic and Plastics Paper:* Vol. 1, pp. 153-54; Vol. 2, pp. 327-28; Vol. 3, pp. 534-37; Vol. 4, pp. 675-79.

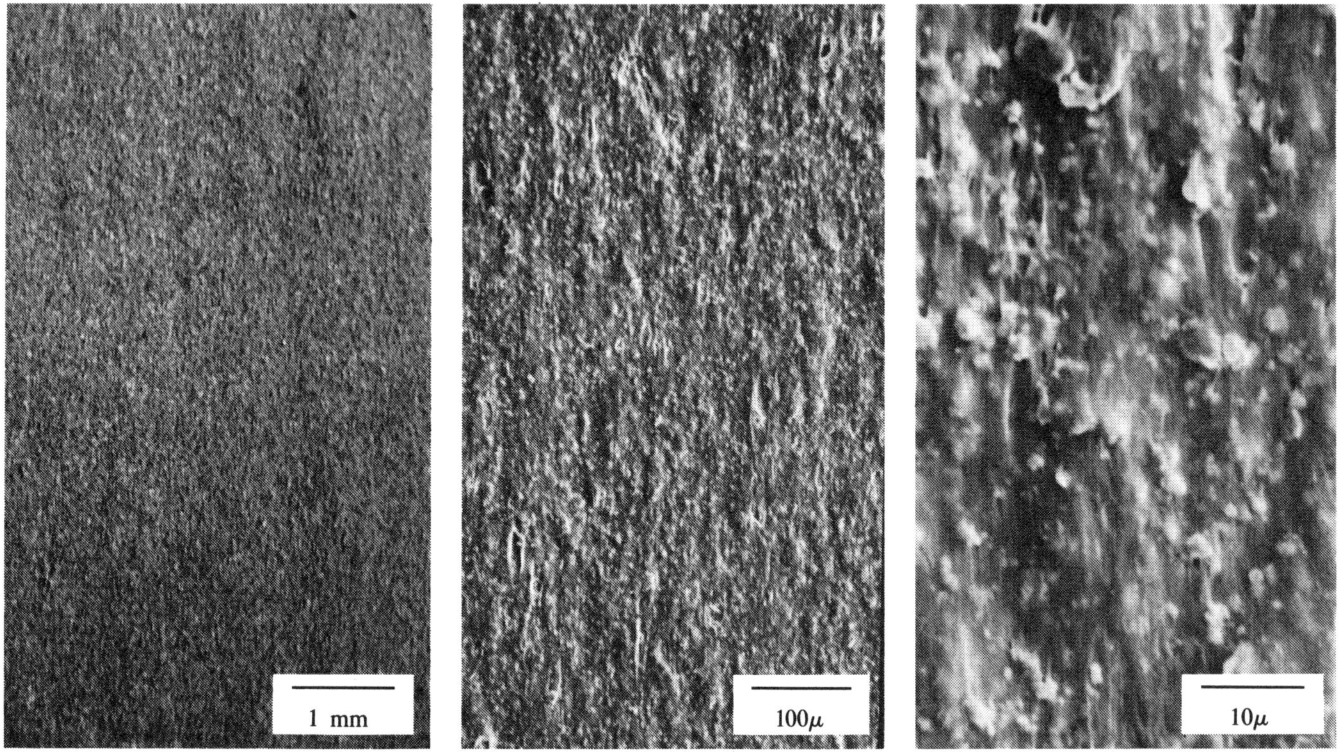

Structure of Yupo FP, a three-ply polypropylene paper, is shown in photomicrographs at: left, 20x; center, 200x; right, 2000x. Manufacturer is Oji-Yuka of Japan. Kimberly Clark has exclusive marketing rights for North and Central America for Oji-Yuka PP durable materials. Courtesy Oji-Yuka.

cooperation between leading firms of the synthetic polymer industry and conventional paper industry.

The polypropylene synthetic paper process, developed by Oji-Yuka Synthetic Paper Research Institute (founded in May 1969) has the following features:

- ☐ Material: filled polypropylene.
- ☐ One-step paperization process from PP pellets to finished paper, ready for printing.
- ☐ Process suitable for large-scale and high-speed production.
- ☐ Process enabling the production of a very wide range of thicknesses of synthetic paper, including paper with a thickness as low as 40 microns.

The integral part of this process is special biaxial orientation of the film by modified tenter method. One of the main features of this paper is its low density of 0.7-0.8; the main applications of this paper's various types are art paper, coated paper, high-quality paper.

According to Oji-Yuka, the printability of this synthetic paper is equivalent to that of coated paper; offset, relief, gravure, and flexographic printing can be applied, using conventional printing ink. The Oji-Yuka Research Institute has also been doing R & D work on synthetic paper based on polystyrene film, and on synthetic paper based on synthetic fibers.

The company's 6,000 ton/yr production capacity plant was completed early in 1972. This paper, which is a filler-contained multilayer product, is supplied in a range of thicknesses between 70 and 120 microns (weight/m^2, 50-90 g respectively). Table 3.13 compares the properties of Oji-Yuka's paper with conventional (cellulosic pulp) coated paper. According to this table, 100-micron Oji-Yuka paper weighs on a square meter basis about 10% less than 80-micron conventional coated paper.

Mitsubishi Petrochemical Company Ltd., a co-owner of Oji-Yuka, has patented the method of "production of synthetic papers."[21] According to this patent, a method has been developed to produce synthetic papers of laminate structure, wherein cut-off edge parts can be utilized for reclamation of the base-film material.

> The paper-like material layer is prepared by extruding the mixture of a thermoplastic resin and an inorganic filler through the central part of a film-forming die, and simultaneously and confluently

21. British Patent 1,323,758, Complete Specification published July 18, 1973.

Table 3.13 *Comparison of OYU-Paper with Conventional Coated Paper*

Properties	*Unit*	*Conventional coated paper*	*OYU-paper (Oji-Yuka)*
Thickness	micron	80	100
Density		1.05	0.75
Whiteness	%	82	89
Opaque ratio	%	90	95
Tensile strength			
MD	kg	5.8	11.3
TD	kg	3.5	24.5
Tensile elongation			
MD	%	2-3	53
TD	%	5-7	25
Tear strength			
MD	kg/cm	26	145
TD	kg/cm	31	87
Folding endurance	times	25	>150,000
Heat seal	kg/2 cm	–	4.5
Reflection ratio of printed paper	%	2.4	2.1
Opaque ratio after printing	%	82	92
Printing surface gloss	%	58	56

Source: Oji-Yuka Synthetic Paper Company, Tokyo, Japan.

Note: The name of the paper produced by Oji-Yuka is now Yupo FP.

> extruding the resin substantially free of an inorganic filler through the outer side parts of the die, thereby to form the edge regions.
>
> The laminated structure is drawn by at least 1.3 times in its longitudinal direction in order to obtain a composite structure, and the resulting composite structure is then drawn under heat by at least 2.5 times in its transverse direction in order to obtain a laminated structure which consists of one biaxially oriented film and of at least one uniaxially oriented film.

Another development by Mitsubishi Petrochemical Co. Ltd., referring to improved foldability combined with good writability and good printability of synthetic papers, has been patented under the title, "Synthetic Papers."[22] According to this patent, synthetic papers with a specific content of a homo- or co-polymer of styrene and/or of a styrene as an agent to improve the foldability, show good results and have "excellent stiffness and tensile strength." The good writability and printability are due to some extent to the presence of microvoids in the

22. British Patent 1,317,911, Complete Specification published May 23, 1973.

films, resulting from the drawing or stretching of the filler-containing film. The patent refers to two types of synthetic papers that are the result of this invention.

1. Nonlaminated structures comprising solely a writable or "papery (paper-like) film."
2. Laminated structures comprising a writable film adhering to at least one surface of a backing film.

According to this patent, synthetic paper produced in this way showed greatly improved foldability compared with biaxially-oriented polypropylene film that contained an inorganic filler but did not contain polystyrene.

The British patent "Production of Synthetic Paper,"[23] granted to Oji Yuka Goseishi Kenkyujo, Tokyo, refers to synthetic paper based on polystyrene film and having a matte surface.

The invention provides in one embodiment a process for producing a film having a matte surface. The process comprises drawing a film made up of a polystyrene resin in which particles of an elastomeric component are dispersed; treating the film thus drawn with a solvent exhibiting dissolving action with respect to the polystyrene resin while exhibiting swelling-inducing action with respect to the elastomeric component; and subjecting the resulting film to conditions that will stop the actions, with respect to the film of the solvent adhering thereto.

This patent specification includes 15 examples referring to the basic films. Most of the examples list films with over 90% content of polystyrene, the remaining content being butadiene. However, example 14 refers to a biaxially drawn film, composed essentially of 50 parts of high-impact polystyrene and 50 parts of HD PE. Example 15 refers to a biaxially drawn film of ABS resin containing elastomeric particles and treated with toluene which forms a toluene layer.

The role of filler particles in synthetic papers and the question of achieving "cavitation-during-orientation" of plastics film, was discussed in a joint paper by J.R. Quick, C.J. Benning, and J.P. Tonelli (International Paper Company, USA) presented at the ANTEC Conference in Montreal.[24] According to this presentation the role of the filler particles was more obvious in the micrographs of the Yupo FP synthetic paper than in the case of Polyart and Acroart. The authors stated that

23. British Patent Specification 1,329,598, published September 12, 1973.

24. Quick, Benning, and Tonelli, "Generation of Microphones in Plastic Film Papers and Processes" (Presentation at the ANTEC Conference, organized by the Society of Plastics Engineers, USA, and held at Montreal, Canada, May 7-10, 1973).

Yupo FP is prepared by stretching a polypropylene film filled with clay and titanium dioxide, and that

> during the stretching process, cavities develop around the filler particles because they cannot stretch and the resin-to-filler bonds are relatively weak. If the resin were strongly bonded to the filler there would be less cavitation and the film would be oriented to a much higher degree around the filler particles. This would not provide the porosity, opacity, and low density needed in a paper substitute.
>
> The YUPO FP paper is made up of three layers. All are filled polypropylene, but the center layer is biaxially oriented and has a lower filler content than the outer layers, which are uniaxially oriented. This structure gives the product balanced physical properties due to the biaxially oriented core, but the surfaces have a high level of porosity and the porosity is more uniform than it would be if the surface layers were biaxially oriented.

The joint paper "The Printability of Synthetic and Plastics Paper" by the late Dr. A. B. Truman and F. D. Munday (Pira, England) presented at the international symposium in Cambridge (September 1973) described the multilayer product as a significant advance in the field of synthetic papers. The preprint of this presentation said that whereas most synthetic papers have a comparatively low stiffness, the stiffness of this multilayer material was as high as that of conventional coated papers.[25]

Early in 1976, Kimberly-Clark (USA), which has been importing Yupo FP synthetic paper from Japan and converting and marketing it under the trade name Kimdura, was still evaluating the possibility of establishing a production unit in the USA.

The latest developments and market assessments of Kimdura were described by Dr. Norman J. Beckman and Richard C. Abb.[26] Dr. Beckman stressed Kimdura's low density of 0.77 g/cm^3 as against a conventional coated paper density of 1.27 g/cm^3, resulting in high yield in ft^2/lb of Kimdura. Kimdura is a 3-layer PP laminate containing inorganic fillers; the extruded core layer is heat stretched in the longitudinal direction for strength development. After lamination of the two surface layers to the core the laminate is heat oriented in the transverse direction. Because the pigment content is higher in the surface layers, a high

25. International Fundamental Research Symposium (organized by Technical Section, BPBMA, London) held at Cambridge, England, September 1973.

26. Beckman at the TAPPI Paper Synthetics Conference, Pittsburgh, October 6-8, 1975 and Abb at the TAPPI International Synthetic Papers & Pulps Symposium, Washington D.C., March 31-April 2, 1976.

Table 3.14 *The Five Basic Grades of Kimdura*

Grade	*Thickness (mils)*	*Yield (ft^2/lb)*
FP-80	3.2	79
FP-95	3.8	67
FP-110	4.5	58
FP-130	5.2	49
FP-150	5.9	42

Source: Norman J. Beckman, Kimberly-Clark Corporation (paper presented at the 1975 TAPPI Paper Synthetics Conference, Pittsburgh, Pennsylvania, USA, October 6-8, 1975).

Note: There is also a special label grade SG.

degree of opacity results.[27] The five basic grades of Kimdura are shown in Table 3.14.

Kimdura has been printed mainly on sheet-fed offset presses, but also by rotogravure and flexographic processes as well as by silk screen printing methods. Specially formulated inks have been developed for Kimdura printing. Richard C. Abb referred to some main markets and potential applications of Kimdura, including labels (e.g., Fasson selected 100% Minute Maid Lemon as its ''Package of the Year'' with its Syntique label made of Kimdura FPG 80) and packaging, with emphasis on laminated products.

Whereas in Japan Yupo FP has been cleared by the authorities as a suitable material for packaging all food products, in the USA Kimdura has at present a limited clearance for dry food and cereals packaging. The clearance by the FDA for all food packaging was delayed because of recent changes in the formulations of Kimdura involving different antioxidants and antistatics, but Abb expressed the hope that the FDA clearance will be forthcoming.

The development of reinforced Kimdura products indicates a very good potential market in flexible wall coverings. The translucent (transparentized) grade of Kimdura has a potential market in engineering applications, including in tracing vellums. Children's books, games, wall charts, maps, etc., where durability and high-class printing are required, are other potential markets. In contrast to Japan, where high-styled reusable shopping bags had a large application, in the USA disposable throw-away kraft bags limited a similar use of Kimdura.

Abb referred to the latest Kimdura product developments, including

27. U.S. Patent No. 3,741,841 contains basic description of the production process.

those that make Kimdura more suitable for use with adhesives and provide better ink adhesion to the smooth Kimdura surface. The grade SG80 has one clear (smooth) surface and one print surface. Other developments include changes in the core and surface-filler ingredients to improve resistance to UV degradation and discoloration. Kimdura Plus, a laminate of Kimdura and Reemay spunbonded polyester material (manufactured by Du Pont), resulting in a sturdy material with very high class print properties, and Kimdura Duo (10-mil laminate of two layers of Kimdura) are among new product developments with interesting market possibilities.

Filled Polyallomer Film

Scott Plastic Paper

Scott Plastic Paper, developed by Scott Graphics Inc., is, in accordance with the classifications of this book, a synthetic paper. It is a film-paper based on titanium dioxide filled polyallomer—a block copolymer of ethylene and propylene.[28]

The linen grade of this paper is embossed on one side and has a normal polymer crystallization surface structure on the other side. The antique grade, which has not been embossed, shows a stronger crystalline structure on one side and a smoother surface on the other side.

Mineral-filled Plastics Films

The Kokusaku Pulp Industry Co. Ltd. (Japan) has developed a synthetic paper based on HD PE film. This synthetic paper, of matte-type and of relatively low smoothness, is a plastics film with specific powdery material adhered to its surface that makes it suitable for writing and printing.[29]

This patented invention also provides a method for producing a composite thermoplastic synthetic resin film. The method comprises extruding thermoplastic resin into a continuous base film in a predetermined path and pressing an advanced sheetlike carrier having one matted surface on which particles of a solid coating material are deposited. The base film is in an unsolidified state to achieve an embedding of the solid material (talc, clay, or pulp powder) on the surface of the film.

28. See Wright and Walbaum, "General Survey of Synthetic Paper Films," *Man-made Polymers in Papermaking: Proceedings of the IUPAC-EUCEPA Symposium, Helsinki, June 1972*, p. 337.

29. British Patent Specification 1,173,778, published December 10, 1969.

Examples in the patent specification include those based on HD PE and on PP films, and one of the examples (based on HD PE) refers to equipment for producing a composite thermoplastic synthetic resin film having a laminated structure in which a resinous base film is sandwiched between a kraft-paper layer and a clay-particle layer.

Kokusaku Pulp Industry together with Sanyo Pulp Co. of Tokyo has formed a joint company—Sanyo-Kokusaku Pulp Company Ltd., Tokyo. Sanyo's developments in the field of speciality synthetic papers have been based on cellular-structure products. The company's speciality synthetic papers have been developed on the base of polyolefines or PVC and have low density and high opacity, based on their fine cellular structure as well as "fairly" high strength, in spite of their cellular structure.[30]

On the whole, cellular structure and oriented film offer advantages in the production of synthetic paper because of the increased stiffness of the film, which improves the printability characteristics of the material, and because less material is used for the required thickness of the synthetic paper, which results in lowering the production cost.

HD PE films filled with Kaolin and a small percentage of titanium dioxide *before extrusion* are another development of mineral-filled plastics films.[31] Although, strictly speaking, they represent a further development of plastics paper, they are described in this chapter because they aim mainly at the graphic industry's market.

Successful trials have been carried out by some polymer manufacturers (e.g., Hoechst, using its modified type of GF 4760 grade of HD PE polymer) and also by some manufacturers of extrusion equipment (e.g., Troester). The same extrusion equipment was used as was used for unfilled film, and the amount of Kaolin, depending on the type of the end product, constituted up to 30% of the weight of the polymer. Among potential applications for this type of plastics paper are visiting cards, calendars, labels, and punch cards.

HD PE/Polystyrene Compound Film

Chemische Werke Hüls (Fed Rep Germany) has developed a range of plastics and synthetic papers made from Vestolen A (high-density polyethylene) and Vestyron (impact-resistant polystyrene), with the application of Litex (styrene copolymers) and of polymer compounds. Samples of these products were exhibited at Interpack, Düsseldorf, May 1973.

30. See Wolpert, *Synthetic and Plastics Paper: Vol. 2, pp. 324-26 and 331.*

31. See Wolpert, *Synthetic and Plastics Paper:* Vol. 3, p. 405 and Vol. 4, pp. 651-52.

These products are suitable for all printing processes (flexographic, rotogravure, letterpress, and offset), and the two-color samples that were exhibited indicated that these products were aiming at the graphic industry's markets. The polyolefins are manufactured by Vestolen GmbH and marketed by Hüls.

It is felt that compounds of HD PE and PS offer some advantages both in processing and in end products. The film is stiffer and can be made either by lay-flat tube or flat-die extrusion or by calendering. The investment cost of a calender is higher than that of an extruder. However, the much-higher production rate in the calendering technique results in lowering operating costs. The surface of the film can be influenced by the choice of the calendering rolls. According to Hüls, the calendering technique offers further advantages in respect of the thickness tolerances and the width of the film.

In connection with the calendering technique, reference is made to the "Calendrette" equipment by Kleinewefers of Fed Rep Germany. The equipment consists of a flat-sheet die and a calender. Francis Shaw & Co., Ltd. (UK), using the Kleinewefers principle, has designed an extruder/calender machine. It is reported that the 90-mm extruder has a 130 kg/hr and the 115-mm extruder a 210-kg/hr output. Tables 3.15-3.18 contain data regarding synthetic and plastics paper developed by this technology.

Table 3.15 *Synthetic Paper Based on Vestolen A, Vestyron, and Litex*

Property	*Testing method*	*Unit*	*Value*
Thickness	–	micron	140
Weight by area	–	g/m^2	168
Weight by volume	–	g/cm^3	1.2
Tear resistance, lengthwise/crosswise	DIN 53455	kp/cm^2	168/131
Tear elongation, lengthwise/crosswise	DIN 53455	%	9/5
Tear propagation, lengthwise/crosswise	DIN 53356	kp/mm	1.3/1.7
Opacity	DIN 53146	%	95.8
Remission rate	DIN 53145	%	88.5
Whiteness rate	Tappi		

Notes: Raw material: Base film from Vestolen AX 2330. A product based on Vestolen A (high-density polyethylene) and Vestyron (impact-resistant polystyrene). Printing layer of Litex (styrene copolymer).

Production process: (1) Extrusion of the film (lay flat tube extrusion or flat die extrusion) or calendering, (2) Coating of the film with Litex.

Source: Chemische Werke Hüls, Technical Bulletin, Marl, Fed Rep Germany, 1973.

Table 3.16 *Synthetic Paper Based on Vestyron and Litex*

Property	*Testing method*	*Unit*	*Value*
Thickness	–	micron	135
Weight by area	–	g/m^2	148
Weight by volume	–	g/cm^3	1.1
Tear resistance, lengthwise/crosswise	DIN 53455	kp/cm^2	438/362
Tear elongation, lengthwise/crosswise	DIN 53455	%	50/55
Tear propagation, lengthwise/crosswise	DIN 53356	kp/mm	0.36/0.47
Opacity	DIN 53146	%	91.1
Remission rate	DIN 53145	%	85.5

Notes: Raw material: Film of Vestyron 618-30 (impact-resistant polystyrene), printing layer of Litex (styrene copolymer).

Production process: (1) Extrusion (flat die extrusion), (2) Biaxial orientation of the film, (3) Coating of the film on both sides by Litex.

Source: Chemische Werke Hüls, Technical Bulletin, Marl, Fed Rep Germany, 1973.

Table 3.17 *Plastics Paper Based on Vestolen A*

Property	*Testing method*	*Unit*	*Value*
Thickness	–	micron	120
Weight by area	–	g/m^2	132
Weight by volume	–	g/cm^3	1.1
Tear resistance, lengthwise/crosswise	DIN 53455	kp/cm^2	296/290
Tear elongation, lengthwise/crosswise	DIN 53455	%	870/958
Tear propagation, lengthwise/crosswise	DIN 53356	kp/mm	28.7/37.5
Opacity	DIN 53146	%	90.0
Remission rate	DIN 53154	%	88.5

Notes: Raw material: Vestolen AX 2590. A high-density polyethylene based on Vestolen A 6042, with additives made printable.

Production process: Extrusion (lay flat tube extrusion or flat die extrusion).

Source: Chemische Werke Hüls, Technical Bulletin, Marl, Fed Rep Germany, 1973.

Table 3.18 *Synthetic Paper Based on Vestolen A and Vestyron*

Property	*Testing method*	*Unit*	*Value*
Thickness	–	micron	120
Weight by area	–	g/m^2	132
Weight by volume	–	g/cm^3	1.1
Tear resistance, lengthwise/crosswise	DIN 53455	kp/cm^2	154/136
Tear elongation, lengthwise/crosswise	DIN 53455	%	36/4
Tear propagation, lengthwise/crosswise	DIN 53356	kp/mm	2.3/3.9
Opacity	DIN 53146	%	89.1
Remission rate	DIN 53145	%	89.5
Whiteness rate	according to Tappi		

Notes: Raw material: Vestolen AX 2330. A product based on Vestolen A and Vestyron, made printable for all printing processes by additives.

Production process: Extrusion (lay flat tube extrusion or flat die extrusion) or calendering.

Source: Chemische Werke Hüls, Technical Bulletin, Marl, Fed Rep Germany, 1973.

Arval

The advantage of adding polystyrene (plus possibly some other additives) to polyethylene has been utilized by La Cellophane, S.A., Paris, in the development of Arval synthetic paper (formerly named Polyane), which is supplied in various thicknesses and sizes:

200 micron of 192 g/m^2
300 micron of 288 g/m^2
450 micron of 432 g/m^2

Oriented PP-Film-Based Synthetic Paper

Pryphane

La Cellophane has also been supplying biaxially oriented polypropylene film Pryphane in various types. One type, SCC, is heat-sealable and has good printability characteristics because of the modification of the surface characteristics. This film is supplied in various thicknesses from 20 microns upward.

La Cellophane's developments in this field were patented, and the patent specification stresses the point that whereas polypropylene films, particularly biaxially oriented films, have a considerable growth rate in packaging and other markets, the use of PP films has still been limited by their shortcomings in heat-sealability and in the characteristics of not being "readily printable."[32] It also stresses that these shortcomings can be remedied by applying heat-sealable coatings and coatings which are more receptive to inks.

This La Cellophane invention includes a method of manufacturing a coated film with improved anchoring properties for coatings (isotactic polypropylene is mixed with between 1% and 20% by weight of a copolymer of ethylene and vinyl acetate in which the vinyl acetate content of the copolymer is between 10% and 50% by weight), extruding the mixture through a slot die to form a film, solidifying the film by cooling, biaxially orientating the film, and then coating it with an anchoring coating composition and/or a heat-sealable or other coating composition.

The examples listed in this patent specification show that PP films produced in accordance with this invention and coated with various products have sealing strengths that are more than double those of coated films consisting wholly of PP. If an anchoring layer is applied to polypropylene films manufactured in accordance with this invention, it is possible to obtain an anchorage for coatings that is much greater than can be obtained with the same anchoring agents and coatings applied to a film wholly of polypropylene.

Newcomers

The confidence in plastics film-based synthetic papers can be seen from the fact that leading plastics manufacturers have been actively engaged in R & D work in this field.

In 1974 Hercules Inc., the Film Division (USA), announced the development of the oriented polypropylene WB 503 film, which in addition to its barrier properties and other PP-oriented film properties, including low specific weight, is white and opaque and has a special surface appearance.[33] Two types of this film were developed, and the 1.5-mil film (about 36 microns) has been finding application for printed labels on cans replacing paper labels. This film, which was offered in up to 60-inch widths, could find a market in a range of applications, including wallpaper, packaging, etc. Tables 3.19 and 3.20 list the typical physical properties of type B of this new synthetic paper, and the physical properties of type B (2 mils = 48 microns) as determined by paper tests.

32. British Patent 1,317,808 "Improvements in or Relating to Polypropylene Films," Complete Specification published May 23, 1973.

33. Hercules Technical Bulletin, "Film Developments," January 30, 1974.

Table 3.19 *Typical Physical Properties for Type B*

Thickness (mil) (range)	1.0-2.0
Yield (in^2/lb) approximately 1.0 mil	53,000
Tensile strength (psi)	12-13,000
Elongation (%)	30-40
Modulus (psi)	110-180,000
Elmendorf tear strength (g/mil) (tear propagation)	4-6
Graves tear strength (lb/mil) (tear initiation)	0.6-0.8
Dart impact strength (ft . lb/mil)	0.6
Opacity (TAPPI, %)	70-80
MVTR (g/100 in^2/24 hr/mil) (100°F, 90% RH)	1.0

Source: Hercules Inc., Technical Bulletin, "Film Developments," Wilmington, Delaware, USA, January 30, 1974.

In 1976 Hercules started marketing type B of this synthetic paper under the trade mark OXWB and reported that it was replacing clay-coated kraft paper. OXWB laminated to PP film finds application in making bags for potato chips sold from vending machines.

SYNTHETIC PAPERS BASED ON SYNTHETIC FIBERS

Synthetic papers based on synthetic fibers (mainly consisting of cellulosic fibers plus a percentage of synthetic fibers) are older than

Table 3.20 *Physical Properties as Determined by Paper Tests*

	Type B – 2 mil
Tensile strength (lb/in)	8-12
Elongation (%)	20-50
Tear strength (g/sheet)	8-10
Fold endurance (cycles) (MIT)	10,000
Burst strength (psi) (Mullens)	40
Clark stiffness (cm^3)	24
Opacity (TAPPI, %)	89
Brightness (GE, %)	73
Basis weight (lb/ream)	18-20

Source: Hercules Inc., Technical Bulletin, "Film Developments," Wilmington, Delaware, USA, January 30, 1974.

those based on plastics films.[34] Synthetic papers based on synthetic fibers and spunbonded materials are more akin to conventional paper—both kinds of paper being basically a sheet consisting of fibers—than to synthetic papers based on plastics film.

Synthetic papers based on synthetic fibers can be "tailor-made" by the choice of the content (ratio of individual components, including synthetic fibers and cellulose fibers), length and thickness of fibers, etc. According to some evaluations synthetic papers based on synthetic fibers are the most superior products in the line from wood-containing to wood-free paper and from wood-free to rag paper.[35]

Synthetic papers based on synthetic fibers, those based on plastics films, and conventional paper all compete in many cases for the same applications. The right choice of material (type of paper) depends on the required properties of the end product, and the production costs of the various materials have to be taken into account in this decision. Each of these materials is superior to other materials (in quality and/or price) in some applications and inferior in others. The penetration of the printing- and writing-paper markets by various types of synthetic paper has been confined to speciality products.

Viscosuisse Emmenbrücke (Switzerland) has been active since 1956 in research on the use of synthetic polymer materials in the paper industry. The Complete Specifications of the company's British patents were published in 1964. Their titles are:[36]

Improvements in the Production of Paper (Patent Specification No. 948,749)
Improvements in the Production of Paper (Patent Specification No. 948,750)
Improvements in or Relating to the Manufacture of Paper-like Materials Comprising Synthetic Fibers (Patent Specification No. 958,430)

The latter patent refers to improved dispersions of synthetic fibers having a reduced tendency to flocculate. These dispersions can be

34. See D. McLean Wyllie, "Review of European Developments" in *Synthetic Fibers in Papermaking*, ed. O.A. Battista (New York: John Wiley & Sons, Inc., Interscience Publishers, 1964). This review includes reference to polyester-fiber paper by Faserprodukte GmbH, Fed Rep Germany, and says that "undoubtedly, one of the most successful European papers comes from the Sihl Paper Mill, Zurich, Switzerland, under the name of Syntosil." The review also states that "as an example of a simple binder, chemically related to the synthetic fiber, Portals Ltd. .... have a patent that may be quoted in which the methyl methoxy derivative of nylon is used with cellulose and nylon fiber."

35. *Allgemeine Papier Rundschau*, Fed Rep Germany, (April 1972), p. 1633.

36. The first two specifications were published on February 5, 1964, the third on May 21, 1964.

Table 3.21 *Properties of Papers Containing Synthetic Fibers and Different Binders*

Binder	*Percent nylon fiber*	*Breaking length in meters (MD)*	
		Dry	*Wet*
SVS binder	40	4,710	2,750
	80	4,480	2,760
Acrylic binder	40	1,515	452
	80	990	495

Source: Viscosuisse, Emmensbrücke, Switzerland.

obtained by coating the fibers with a water-soluble N-ethoxylated polyamide. This invention provides a process for the production of paper-like material comprising synthetic fibers by forming the paper-like material from an aqueous suspension containing these fibers, which have a coating of a water-soluble N-ethoxylated polyamide.

Viscosuisse stressed the point that it is of special importance that papers containing both natural and synthetic fibers use a binder with a good affinity to a substrata (e.g., cellulose, nylon, polyester, etc.) since the tensile strength and other properties of paper containing synthetic fibers depend to a large extent on the quality of the bonding agent. Viscosuisse has developed such binders. Table 3.21 shows the improvement achieved in paper containing 40% and 80% nylon fiber when using Viscosuisse's special SVS binder instead of the conventional acrylic bonding agent.

In the endeavor to commercialize Viscosuisse processes it has been found more advantageous—from both the technical and the economic point of view—to produce speciality papers containing a blend of natural and synthetic fibers than to manufacture a 100% synthetic fiber paper. By variation of the components and ratio of natural fibers, synthetic fibers, and specially developed synthetic binders, it is possible to produce a wide range of papers of different properties. It is noteworthy that similar conclusions were recently reached in the case of synthetic pulp applications, namely, that blends of synthetic pulp and cellulose are producing speciality papers of a bigger potential than 100% synthetic pulp paper.

Syntosil and Artosil

SIHL, Zurich Paper Mill on Sihl (Switzerland), has commercialized the Viscosuisse developments by manufacturing various types of Syntosil and Artosil synthetic papers. It is claimed that these papers, which are produced on conventional Fourdrinier machines, have considerable technical advantages over latex-coated papers and have mechanical strengths superior to those of prime-quality rag paper.

Syntosil is water-, tear-, abrasion-, and UV-resistant. It is supplied in various types, depending on the composition of raw materials, with weights from 45 to 350 g/m². It can be printed by offset, letterpress, and screen printing processes. This hygroscopic and elastic material consists partly of synthetic ingredients that are harder than cellulose fibers. It therefore requires more ink and a higher printing pressure in letterpress printing than conventional paper, but it has the advantage of deep ink penetration resulting in a very high abrasion resistance.

Syntosil has been finding application in the manufacture of geographical maps, since, according to SIHL, these maps can be folded more than 20,000 times without affecting the readability even at the folds. Other applications include license documents, instruction tables, forms, tags and labels, outdoor posters, etc. Syntosil is particularly useful when high water and tear resistance are required.

According to SIHL, type 61 (modified type 60) Syntosil has been used for printing bank notes for various countries. Syntosil also finds application in punched cards for machine tools. Perforag Ltd. (Switzerland) has been supplying Long Life Tape Syntosil, designed for numerically controlled machines with tape input, including machine tools as well as office machinery. This tape is available in various colors, is antistatic treated, and is claimed to have an excellent reproduction at a high repetitive rate, an opacity of 99%, and a minimum wear of punches, etc. Some clients have been using Artosil for punched data cards for business machines. In these applications synthetic papers based on synthetic fibers compete with conventional paper and also with some plastics (e.g., polyester films).[37]

In 1971, SIHL introduced Artosil, which is cheaper than Syntosil and fills the gap between conventional paper and Syntosil. Artosil type 50 of 100-g/m² weight has a specific weight of 0.9 g/cm³. The folding resistance of Artosil-50 in dry conditions is about one-third that of Syntosil-60, but is about five times higher than that of ordinary map paper. Like conventional paper, it can be processed on folding machines, is water-resistant, and finds applications in the same fields as Syntosil. Offset, letterpress, and screen printing processes can be applied in the case of Artosil.

37. In some cases plastics film laminated on both sides with paper has been used for electronic data processing cards. Papierindustrie Döbbelin & Boeder, Fed Rep Germany, has been supplying this material. The company also supplies punched tape products made from plastics film, from plastics impregnated paper, and also from material which includes synthetic fibers. The range of these products indicates the competition and cooperation of the conventional paper and synthetic polymer industries.

Table 3.22 *Weights and Volumes of Syntosil and Artosil*

	Specific weight	*Specific volume*
Syntosil type 100	0.8 g/cm^3	1.3 cm^3/g
Syntosil type 60	0.8 g/cm^3	1.2 cm^3/g
Artosil type 50	0.9 g/cm^3	1.1 cm^3/g

Source: SIHL, Zurich Paper Mill on Sihl, Zurich Switzerland.

Note: Syntosil type 60 and type 100 and Artosil type 50 have a weight of 100 g/m^2.

According to SIHL, the approximate price ratios (based on prices for 1 MT ex-works) for Syntosil, Artosil, and woodfree map paper have been as follows:

Syntosil type 100: about six times more expensive than ordinary woodfree map paper.
Syntosil type 60: about four times more expensive than ordinary woodfree map paper.
Artosil type 50: about twice as expensive as ordinary woodfree map paper.

The approximate thickness of Syntosil and Artosil of 100-g/m^2 weight is 0.12 mm. Specific weights and volumes are shown in Table 3.22. According to SIHL, the strength properties of Artosil type 50, Syntosil type 60 and type 100, and ordinary map paper compare as is shown in Table 3.23.

One of the main applications of these synthetic papers, with their water resistance and high folding endurance, is for outdoor maps. In some cases, however, it has been found that the quality of these papers exceeds the requirements. Detailed topographical maps usually have to be reissued before the old ones are worn out, which means that a lower quality at a cheaper price "would do the job."

Neobond and Pretex

Faserprodukte GmbH of Lahnstein (Fed Rep Germany), which belongs to the Feldmühle concern of Düsseldorf, manufactures Neobond and Pretex. Neobond (formerly known as Silbond), which the company describes as a synthetic paper since it is made mainly from synthetic

Table 3.23 *Strength Properties Compared*

	Artosil Type 50	Syntosil Type 60	Syntosil Type 100	Ordinary map paper
Tensile strength				
Dry: Lengthwise (m)	6,900	7,600	6,600	8,700
Crosswise (m)	4,100	5,200	4,400	4,400
Wet: Lengthwise (m)	2,600	4,000	3,900	600
Crosswise (m)	1,500	2,700	2,700	300
Elongation at rupture				
Dry: Lengthwise (%)	4.5	5.0	6.0	2.6
Crosswise (%)	9.5	8.5	13.0	8.0

Source: SIHL, Zurich Paper Mill on Sihl, Zurich, Switzerland.

fibers, is supplied in a range of weights from 100 g/m^2 to 220 g/m^2. Pretex, which the company describes as a plastics paper, has a small content of synthetic fibers and is plastics impregnated. It is supplied in a range of weights from 100 g/m^2 to 250 g/m^2. According to Faserprodukte, compared with conventional coated paper, Neobond has many far superior characteristics in tear-, water-, and chemical-resistance, dimensional stability, etc.; it also has a very high double-fold capacity.

Pretex fills the gap between Neobond and conventional paper and is cheaper than Neobond. Both products can be printed by offset, letterpress, and screen printing processes, but quick-drying inks have to be used. The main applications of both products are maps, book jackets, board mountings, posters, tags and labels, catalogues, and menus. In addition Neobond is used for drivers' licenses, official documents, etc.

SPUNBONDED AND NOMEX PRODUCTS

Spunbonded products are continuous-filament nonwoven structures produced by a method that integrates the spinning of fibers with the formation of sheet products. According to Du Pont, the continuous-filament character of the product leads to higher tensile and tear strengths as well as to greater toughness compared with "short fiber" nonwovens and many woven fabrics.

Spunbonded products fall into the category of products between paper and film on the one hand and woven fabrics on the other. They offer new properties and opportunities for converters to make a wide range of new end products, from very lightweight flexible structures to heavy stiff materials. Some of these new products have the appearance characteristics of paper with very high opacity per unit of weight. At the same time they share the advantages of synthetics by being resistant to water and to the degrading effects of rot, of mildew, and of insect attack.

E.I. Du Pont de Nemours & Co. Inc. (USA), introduced in 1965 three types of spunbonded products; since then these products have been further developed, overcoming problems of their further processing and use. At the same time, new interesting applications have been found for an increased range of these products.

Du Pont's three spunbonded products are:

Typar: spunbonded polypropylene
Reemay: spunbonded polyester
Tyvek: spunbonded linear polyethylene

To this group of spunbonded products Du Pont has added Nomex nylon, which is not a spunbonded product but which is available in the form of paper sheet as well as in filament yarn and staple fiber.

The main applications of Typar and Reemay are in the textile fields. However, Reemay finds some application in coated and laminated structures, including filtration and electrical uses. Typar, which was developed for primary backing of tufted carpets (competing with jute and woven polypropylene tape fabrics), also finds application in heavy packaging, where in some cases it is laminated to an asphalt impregnated kraft paper.

Tyvek, on the other hand, has been finding increasing use in the paper industry's markets.

Nomex

Nomex nylon has the strength and toughness of Nylon 6.6. In addition it has the following superior characteristics compared with Nylon 6.6:

- ☐ Outstanding thermal and chemical stability. (At 250°C, the melting point of Nylon 6.6, Nomex still retains 60% of its room-temperature tensile strength.)
- ☐ Will not melt or support combustion. (At temperatures above 500°C Nomex will char and can be consumed by fire, but it will not continue to burn if the ignition source is removed.)
- ☐ Initial modulus (resistance to deformation) about four times as great as Nylon 6.6

Table 3.24 *Typical Physical Properties of Nomex Type 410 Nylon Paper*

	Thickness (mils)					*ASTM Test Number*
	2	*5*	*10*	*20*	*30*	
Tensile strength (lb/in)						
MD	22	70	170	370	550	D-828-60
TD	13	45	100	250	380	D-828-60
Elongation (%)						
MD	10	16	24	25	25	D-828-60
TD	8	14	20	21	22	D-828-60
Elmendorf tear (g)						
MD	95	260	550	1,100	2,000	D-689
TD	140	450	900	2,100	2,700	D-689
Finch edge tear (lb/in)						
MD	20	70	130	160	220	D-827-47
TD	10	25	60	60	100	D-827-47
Shrinkage at 300°C (%)						
MD	1.6	0.9	1.5	1.4	1.4	
TD	1.9	1.3	1.0	0.8	0.8	
Thermal conductivity (Btu in/hr · ft^2 · °F)	–	–	0.76	0.78	–	
Basis weight (oz/yd^2)	1.2	3.2	7.3	15	23	

Source: Du Pont, "Nomex Technical Bulletin," Wilmington, Delaware, USA.

Nomex products have high strength and abrasion resistance (high fold resistance in case of Nomex paper), very high resistance to radiation, and a high resistance to most hydrocarbons and other organic solvents (concentrated sulphuric acid and sodium hydroxide degrade Nomex only after 100 hours' exposure above 50°C).

Nomex Paper Properties

Floc (short, uncrimped staple) can be combined with small fibrous binder particles of Nomex polymer (called fibrids) to produce sheet structures by using conventional papermaking techniques. After formation, these two constituents are permanently bonded together by high-temperature calendering without the use of additional binders, fillers, or sizes. Uncalendered paper is also available. Nomex calendered paper is supplied in a range of thicknesses (e.g., type 410 paper from 2-mils to 30-mils thickness). Tables 3.24 and 3.25 show the typical physical and electrical properties of some of these papers.

Volume and surface resistivities at room temperature are higher than 10^{16}, decrease to about 10^{12} at 250°C, and are independent of paper thickness.

Table 3.25 *Typical Electrical Properties of Nomex Type 410 Nylon Paper*

	Thickness (mils)				
	2	*5*	*10*	*20*	*30*
Dielectric strength (v/mil)*					
AC rapid rise	450	600	750	700	600
AC 1-mil hold	–	–	600	–	–
AC 1-hr hold	–	–	500	–	–
DC rapid rise	–	–	1,200	–	–
Dielectric constant (10^3 Hz)†	2.0	2.3	2.6	3.3	3.4
Dissipation factor (10^3 Hz)†	.007	.010	.014	.018	.020

Note: Tables provided by Du Pont also include thicknesses of 3, 7, and 15 mils.

* ASTM D-149, using two-inch-diameter electrodes.
† ASTM D-150, using one-inch-diameter electrodes under 20-psi pressure.

Source: Du Pont, "Nomex Technical Bulletin," Wilmington, Delaware, USA.

Nomex Paper Applications

Nomex paper possesses a unique combination of good electrical properties, a long useful life at temperatures above 220°C, and outstanding mechanical strength, toughness, and resistance to mechanical stress. It is reported to have been widely adopted for use in electrical insulation and in other industrial applications.

Electrical uses of Nomex paper include:

- ☐ Turn, layer, section, lead, and tap insulation for application in transformers.
- ☐ Conductor, coil, slot, phase, wedge, and end insulation in motors and generators.
- ☐ Printed circuit backings, terminal and potentiometer strips, cable insulation, creped and pressure-sensitive tapes.
- ☐ Spiral and convolute-wound tubes, coil bobbins, V-rings, and a variety of other formed parts.

Non-electrical uses include:

- ☐ Mould-release papers.
- ☐ Compressor vanes.
- ☐ Corrugated and honeycomb structures for aircraft and aerospace applications.
- ☐ High-temperature ducts and firewalls.

Nomex paper can also be laminated with another sheet of Nomex paper as well as with other materials to form composites (either flexible or rigid), or can be coated with various resins, for applications where special characteristics are desired.

Tyvek

Tyvek, a spunbonded sheet of 100% linear polyethylene, is produced by bonding, under heating and pressure, a web of continuous fibers ranging from less than 0.5 microns to about 10 microns in thickness (average about 4 microns), and without any external binders, sizing agents, or fillers. The extremely small diameter of the fibers leads to a high opacity and whiteness of the product (in the absence of any filler pigment).

Tyvek is supplied in various types: Tyvek 10 in nominal thicknesses of 5 to 10 mils (nominal weight: 1.3 to 2.7 oz/yd^2 respectively) and Tyvek 14 and 16 in nominal thicknesses of 5 to 7 mils (nominal weight 1.15 to 1.6 oz/yd^2).

There are two different types of bonding:

1. Area: over the entire sheet (produces Tyvek 10).
2. Point: confined to discreet spots separated from one another by unbonded fibers (produces Tyvek 14 and 16).

The finished sheet characteristics and properties are strongly influenced by the bonding system. Table 3.26 compares the properties of area-bonded Tyvek, kraft paper, and cotton sheeting. Bending-length comparisons are shown in Table 3.27.

Tyvek area-bonded sheets have a paper-like appearance, combining high opacity and whiteness with strength, toughness, and tear resistance, and they find a growing market as book-covering material and in a wide range of applications in the graphic arts industry—including textbooks, maps, advertising banners, tags and labels, heavy packaging—and in wall coverings.

Tyvek point-bonded sheets are used primarily in the fields of industrial and consumer apparel and mattress construction. In the packaging field, they find application in making draw-string bags for mechanical parts and other small articles.

It is significant that some distributors of Tyvek in Europe are leading papermaking companies. Du Pont's French agents for Tyvek DHJ (Europe) S.A., Paris, were distributing at the 1971 ITMA exhibition publicity material of textile machinery in folders made from Tyvek.

Table 3.26 *Comparison of Area Bonded Properties*

	Spunbonded PE sheet	*Kraft paper*	*Cotton sheeting 64 x 64*
Basis weight (oz/yd^2)	2.2	2.4	2.7
Thickness (mils)	8.0	4.9	10.2
Strip tensile (lb/in)	46.0	27.9	12.1
Strip elongation (%)	28.0	4.3	12.0
Work-to-break ($in \cdot lb/in^2$)	8.0	0.44	1.7
Tear, Elmendorf (lb)	1.3	0.29	2.9
Mullen burst (lb/in^2)	115	25.2	83.4
Opacity (B&L, %)	0.95	0.96	0.63
CSIA abrasion (cycles)	59	23	18
Smoothness (μin)	80	75	280
Initial modulus ($lb/in^2 \times 10^{-3}$)	120	141	6.5

Source: Du Pont, "Tyvek Technical Bulletin," Wilmington, Delaware, USA (quoted in Wolpert, *Synthetic and Plastics Paper*, Vol. 1, 1971, p. 109).

Note: All properties are averages of machine-direction and transverse-direction values.

In the USA, Tyvek is being used for making envelopes. Table 3.28 contains comparison data of two grades of Tyvek and several types of kraft paper. Tyvek 1056 has a thickness of 6 mils, and Tyvek 1073 has a thickness of 8 mils (1 mil = 0.001 in).

Because of the light weight of Tyvek tough envelopes compared with that of the equivalent paper envelopes, postage savings are achieved; in some cases these savings are greater than the price difference of these envelopes, which means that for the end-user, Tyvek envelopes may in fact be less expensive than some competitive materials.

Tyvek is also used for sterile packaging, where it is reported to have several advantages. Because of its high strength it resists delamination when sterile packages are peeled open, and compared with paper, there is reduced chance of *fuzzing*, or of particles falling into the sterile field.

Table 3.27 *Bending Length Comparisons*

	Woven cotton	*Spunbonded polyethylene*		*Kraft paper*
		Point-bonded	*Area-bonded*	
Basis weight (oz/yd^2)	2.7	2.0	2.2	2.4
Bending length (cm)	2.2	4.5	6.0	7.2

Source: Du Pont, "Tyvek Technical Bulletin," Wilmington, Delaware, USA.

Table 3.28 *Physical Properties of Envelope Kraft Papers and Tyvek Spunbonded Olefin*

Sample	*Breaking strength (lb)*		*Work-to-break* (in · lb/in²)*		*Mullen burst (psi)*	*Elmendorf tear (g)*	
	MD	*TD*	*MD*	*TD*		*MD*	*TD*
Kraft							
20 lb	38.8	15.3	0.40	0.54	40	102	112
24 lb	43.8	20.3	0.51	0.72	37	147	144
28 lb	56.1	19.5	0.62	0.61	57	147	110
32 lb	68.7	24.1	0.75	0.69	57	192	155
40 lb	61.8	24.9	0.74	0.87	80	243	218
Tyvek							
1056	29	30	3.9	5.2	120	410	410
1073	47	48	6.0	7.8	175	450	450

Source: Du Pont, "Tyvek Technical Bulletin," Wilmington, Delaware, USA.

* The work necessary to break the envelope by applying tensile force; a combination of breaking strength and elongation.

According to the Du Pont brochure, "for superior protection against contamination, specify TYVEK spunbonded olefin for sterile packaging." Tyvek is a superior bacteria barrier compared with commonly used sterile packaging paper. Unlike film, Tyvek is easily penetrated by sterilizing gas, even though is is not easily penetrated by bacteria. The Gurley air porosity of Tyvek 1073B with approximately 14 seconds is very low. The tensile strength of Tyvek 1073B at room temperature is about 40 lb/in. Its elongation to break is about 25% in the machine direction and about 30% in the cross direction. The dimensions of this material change less than 0.01% between 0% and 100% RH at constant temperature.

Tyvek was introduced in Europe several years later than in the USA, and lately has been registering increased sales. Wiggins Teape (Mill Sales) Ltd. (UK), which distributes Tyvek in Britain, Eire, Scandinavian countries, and Portugal, reports that the sales of Tyvek have been increasing. Tyvek has been used in Britain for specialized technical manuals and for special clean-room applications. The use of Tyvek for envelopes shows an upward trend. Self-adhesive- and heat-seal-coated Tyvek 10 for tags and labels are reported to be making progress, partly in competition against PVC film. Laminates consisting of Tyvek/paper and Tyvek/LD PE film for sacks used for overseas shipment have been penetrating the market.

Red Bridge Book Cloth Co. Ltd. (UK) has been converting Tyvek by coating it for making bookbinding material.

Appleton Coated Paper Company is among the U.S. paper manufacturers who have been using Tyvek. In Appleton's Ascot-coated products the base material is Tyvek. In some cases Ascot products replace conventional paper (maps and charts, tags, labels, signs, book covers); in other cases they replace cloth. Colored coatings can by applied as well as fluorescent coatings. This material has been used as base for wallpaper and for technical applications. Coatings can be designed for high fidelity four-color offset printing.

Spun-Tuff and Texoprint

Kimberly-Clark Corporation (USA) has been manufacturing Spun-Tuff, which is Tyvek, coated on one side. It is significant that Texoprint (which the company has been marketing for over 20 years) and Spun-Tuff have been used for the same applications although Texoprint is a product made from kraft fibers and polymers whereas Spun-Tuff is made from polymers only. The Production Data Texoprint brochure describes Texoprint as "a durable plastic printing paper, consisting of a balanced combination of strong bleached Kraft fibers and latex plus opacifiers and printing fillers." Bleached kraft fibers are formed on the Fourdrinier into a low-density absorbent sheet which is then saturated with elastomeric polymer latex and opacifying fillers (picking up 60% of its own weight in dry solids). The then dried sheet is coated on both sides with printing fillers, including titanium dioxide and flexible latex binder made from special, nonspecified plastics. It is stressed that latex replaces the normal cellulose bonding material in the fiber base, and that flexible synthetic resins replace conventional starches, glue, or casein binders in the coating. Among applications of Texoprint are cover mounts, flip charts, games, pressure-sensitive labels, children's books, score cards, wall charts, maps, posters, etc.

The 1971 study on synthetic paper by Paperfilm Associates (USA) estimated that in the field of durable book papers the market for latex-impregnated book papers in the USA was estimated to be 6,000 to 8,000 tons. The study added that the best-known products were Texoprint by Kimberly-Clark and Tenselex by Sorg Paper Company.[38] From these observations and comparisons Paperfilm Associates concludes that synthetics will be very strong contenders for the durable-book-paper

38. Sorg Paper Company, Middletown, Ohio, USA, has been supplying Tensalex latex impregnated printing paper in light, medium, heavy, and super-heavy weight grades. The company also manufactures "highly specialised latex saturated papers, some with synthetic fibre and unique latex systems" (Wolpert, *Synthetic and Plastics Paper:* Vol. 4, p. 725).

Table 3.29 *BMI Specifications for Covers, Nonwovens*

Test	*BMI specification*	*Representative S-1079-D*
Breaking load (lb/in)		
MD	40 min	58-60
TD	40 min	60-64
MD + TD	⩾95	118-124
Elongation (%)		
MD	7	25+
TD	10	30+
Elmendorf tear (lb)		
MD	0.46 min	1.1+
TD	0.49 min	1.1+
Fold endurance (flexes)	2,000 min	20,000+

Source: Richard E. Seamans, "Tyvek Spunbonded Olefin – End-Uses." (Paper presented at TAPPI International Synthetic Papers & Pulps Symposium, Washington, D.C., March 31-April 2, 1976.)

Note: Specifications are for as received and after exposure to 105°C for 24 hours.

markets as soon as users develop the experience of efficiently handling the new products.

Book covering, particularly for textbooks, is one of the largest single uses of Tyvek, and about 80% of all elementary and high school textbooks used in the USA are at present bound in coated Tyvek.[39] Table 3.29 gives the major property requirements for school book covers specified by the American Book Manufacturers Institute (BMI), and indicates how a specific Tyvek product meets these standards.

Most textbook covers are printed by sheet-fed offset lithography. The printing is protected by a lacquer overcoat in order to increase durability. Uncoated printed sheet is used, though to a lesser extent, in the binding of trade books. (*Reader's Digest* hardcover condensations have been bound in uncoated Tyvek for several years.)

A highly specialized application of Tyvek is for military packaging in accordance with the Mil-B-131 specification. In this case a four-layered composite material is produced. Tyvek is laminated to aluminum foil with branched polyethylene as the adhesive, and the aluminum surface is provided with a further coat of branched polyethylene to achieve heat sealability. Heat-sealed packages using these laminates can be evacuated, maintaining a vacuum for a long period.

39. Richard E. Seamans, Du Pont, "Tyvek Spunbonded Olefin—End-uses" (Presentation at TAPPI International Synthetic Papers and Pulps Symposium, Washington, D.C., March 31-April 2, 1976).

Docan Spunbonded Material

Dr. R. Hoffmeister, Lurgi Mineröltechnik GmbH, in his presentation "Spunbonded Material Production: Technology and Economics"[40] stressed the point that "spunbonded webs can compete with, or be an accessory to, the paper industry in the widest sense. Spunbonded webs will enter the traditional paper products market, and will assist in the development of new and improved paper products."

Lurgi (Fed Rep Germany) has developed a process and plant for making spunbonded materials. This process consists of thermoplastics granules being melted and extruded into filaments which are oriented and deposited as a random web. This process has been named the *Docan spunbonded material producing method*. The individual filaments are in the range of 1.5 to 15 deniers, and the weight of Docan nonwoven virgin material ranges from 10 to 1,000 g/m^2. The semifinished "virgin" material is further processed to finished products by heat sealing and thermosetting under pressure, bonding, etc., at the clients' plants. The choice of the raw material (PP, HD PE, polyester, nylon, polyisobutadiene), the weight of Docan material, and the method of further processing depend on the end product to be made. Among the variety of applications of this material is a material for lamination with paper.

The Docan process has been licensed to a number of manufacturers in various countries, and the majority of end products made by this process are aimed at textile, floor covering, and industrial applications. Crown Zellerbach Corporation (USA), one of the licensees of this process, is active in the production of the fine grades of spunbonded materials. In addition, the company manufactures laminated-tissue-paper products, including disposables for hospitals. These products, marketed under the trade name Crowntex, have high-absorption characteristics and possess a high wet strength.

Dr. Hoffmeister referred to the fact that the high strength of thermoplastics materials, particularly in wet conditions, is an important factor. There is also a range of products, which have been made from paper, but in which the specific properties of cellulose are not of primary importance. Wall coverings belong to this category. In this field the trend is toward synthetics, and it is significant that some Lurgi licensees have been producing material for wall coverings.

Products like spunbonded scrim laminated between two layers of tissue paper are examples of cooperation between the synthetic polymer and the conventional paper industry resulting in end products with

40. At "Plastics and Paper" Conference (organized by the Plastics Institute), London, May 1973.

Table 3.30 *Properties of Heat-Sealed Polypropylene Spunbonded Webs*

	Unit	*MD*	*TD*
Basic weight 20 g/m² single filament 1.5 den			
Strip strength (sample width 15 mm)	kg	0.8	0.82
	kg/mm²	0.75	0.77
Elongation at break	%	8	9
Basic weight 50 g/m² single filament 3.0 den			
Strip strength (sample width 15 mm)	kg	2.1	1.95
	kg/mm²	0.73	0.68
Elongation at break	%	6	5.5

Source: Dr. R. Hoffmeister, Lurgi Mineralöltechnic GmbH, Frankfurt a/m, Fed Rep Germany (paper presented at the "Plastics and Paper" Conference, London, May 1973).

required functional properties that the raw materials of one industry alone cannot provide. Products used in the medical (surgical) fields are also examples of such cooperation.

All spunbonded webs have two characteristics in common: they consist of continuous filaments, and owing to the random arrangement of these filaments they are isotropic (i.e., they have almost equal properties in both directions). This isotropy is an outstanding advantage of spunbonded webs over the dry-laid and, above all, wet-laid webs. Spunbonded materials have a higher elasticity than wet-laid webs; connected with this, voluminosity and high bulkiness are features of spunbonded materials. Table 3.30 lists the properties of heat-sealed polypropylene spunbonded webs.

The spunbonded technology is a continuous, one-step route from polymer chips to web (avoiding the fiber cutting and the subsequent separation of the cut staple fibers). According to Hoffmeister the conversion cost of polypropylene chips into spunbonded web, 8 denier (single filament) at 80g/yd², is about DM 0.10. These costs include labor, utilities, maintenance, taxes and insurances, depreciation on equipment and buildings, and license fees. But they do not include the cost of such important items as polymer chips, overheads, sales, and sales promotion.

Although the investment required for spunbonded web equipment is higher than that for wet-laid web equipment, Hoffmeister stressed the point that spunbonded material production required less personnel (3,000-4,000 tons/yr equipment can be operated by four men per shift), and that coupled with a higher output of spunbonded web the investment cost related to a ton of web product is of the same magnitude as that for dry and wet processes.

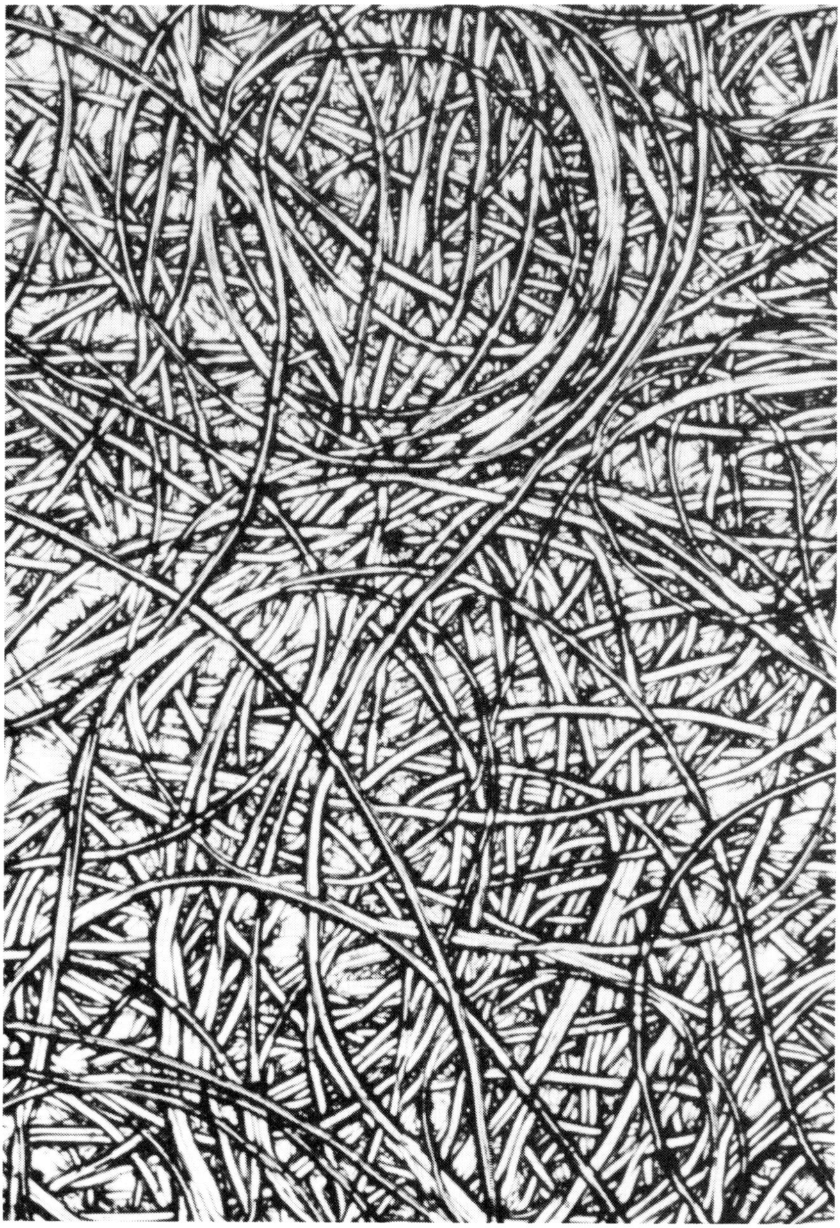

Docan spunbonded materials are made by a process developed by Lurgi Mineröltechnik GmbH, Fed Rep Germany. Semi-finished material is further treated depending on final product. Photo-micrograph shows long fiber matting giving high elasticity and bulk. Courtesy Lurgi Mineröltechnik.

Cerex Spunbonded Material

Monsanto Company (USA) has introduced Cerex, a spunbonded material made from Nylon 6.6 monofilament fibers bonded to each other by a special process; this material does not contain any binders or adhesives. It is supplied in a wide range of weights (and up to 2.9-m widths) and among its various applications is scrim reinforced paper which can be used for making semidisposables (e.g., surgeon dresses). Cerex can also be used for filtration materials, since it does not clog as do paper filters and is considerably cheaper than the corresponding grades of cotton sheets used in filtration.

PRINTING ON SYNTHETIC PAPERS

The different types of synthetic paper, spunbonded materials, and plastics paper, as well as various plastics films, present the printer with specific problems related to the surface of the individual material, to the question of ink/material relation, and to the selection of the most suitable printing process, inks, and printing equipment.

Forward-looking manufacturers of printing inks and printing equipment have been carrying out R & D work (based on their experience with conventional materials) with newly developed materials for the graphic industry. It is to be expected that the growth rate of the new materials replacing conventional pulp/paper products will result in technological developments in the printing of these materials for the graphic industry, and that these developments will in fact be instrumental in accelerating the market penetration by synthetic materials in applications in which their functional advantages over pulp/paper are important.

Plastics films usually have a highly smooth and impermeable surface; specially formulated inks must therefore be used. At the other end of the range of new materials are spunbonded materials such as coated Tyvek and other materials with a porous surface having good ink penetration. Trial runs of printing wall maps for schools on coated Tyvek have been successful.

Film-based synthetic papers are manufactured by various techniques; surface coating, use of fillers, and surface etching are among the methods of converting thermoplastics film into synthetic paper. Corona discharge treatment or chemical etching improve the printability of the product. Experience shows that synthetic papers can be printed satisfactorily as long as the specific properties of the material and printing method have been taken into consideration; for example, specially developed oxidation-drying inks must be used in litho and letterpress printing in order to achieve good drying and keying. In fact, less ink is normally required with these materials than with conventional paper because of their nonabsorptive properties (except in heavily coated materials).

For effective printing, the litho-damper settings should be kept as low as possible. Distilled water is recommended for damping. Anti-setoff spray used very sparingly has been applied without reducing the rub resistance. When printing is done on sheets, it is recommended to stack them in small piles and air them at regular intervals.

According to BXL, manufacturers of Polyart, the ink transfer and drying are of particular significance in litho and letterpress printing, and prints of "excellent definition and good tonal range" are achieved, allowing the use of very fine screens up to 80 lines/cm in halftone reproduction. BXL stresses the point that because of the low absorption of ink into the substrate, degradation of color and loss of intensity are prevented. The company also states that printing by gravure, flexography and silk-screen is not difficult, since such processes are widely applied in printing of plastics for packaging applications.

Some printers have found that the use of powerful drying catalysts (e.g., cobalt catalysts) has various disadvantages, but that the addition of a suitable water-soluble cobalt salt to the fountain solution can be used, since it emulsifies into the ink at the plate and avoids skinning and drying-on problems. However, attention should be paid to the fact that additional catalysts may cause heat generation in the stack, resulting in the ink getting sticky during drying. Some printers have found that when the weather is cool, alcohol additives should be used in the water fount. Winstones of Harefield (UK) is among the suppliers of such alcohol additives.

Synthetic papers that are pigment-coated can be printed with conventional quick-setting inks. In the case of multicolor printing, however, it has been found that sometimes during the first printing the surface of the material is "sealed" and that this prevents the setting and drying of the subsequently superimposed colors (inks).

According to I. Reinbold and H. Lunk, an economic evaluation has demonstrated that pigment coating for the surface improvement of films results in the lowest comparative costs for synthetic printing materials on a polystyrene film base.[41] They feel that among the many important properties of printability the acquisition of a sufficiently high level of opacity is perhaps of greatest significance, and they referred to trials which have shown that, along with the use of titanium dioxide in the coating, two-sided pigmentation of the film is beneficial.

---

41. In *Man-made Polymers in Papermaking: Proceedings of the IUPAC-EUCEPA Symposium, Helsinki, June 1972.*

On the question of *ghosting*, Coates Group of Companies (UK), manufacturers of ink, said that

> remarkable examples of gloss and refusal ghosts have been observed on the reverse of polystyrene/butadiene and more particularly polyolefin based synthetic papers. They are explained by the diffusion of volatile oxidation by-products from print on the first side onto the back of the sheet above in the stack. These vapours are held very strongly and either accelerate or inhibit the drying of an ink subsequently printed on the affected surface—giving a gloss ghost—or modify the wetability of the surface and cause refusal ghosting. The presence of the by-products cannot be detected visually on the sheet, and they are very difficult to remove although warming, e.g., by IR lamps, is said to help. Alternatively pre-printing the affected surface with varnish heavily loaded with extender is often a satisfactory answer.[42]

Coates Group has developed for litho and letterpress processes modified grades of PLF inks for printing on synthetic paper. These modified inks have no mineral-oil content.

Printers who have experience with several kinds of synthetic paper feel that some of these papers are suitable for being printed on both sides, whereas others are preferably used for printing only on one side of the paper (e.g., posters).

Some printers emphasize that there is a critical relationship between the amount of moisture built up in the lithographic process and the surface of synthetic paper. According to these printers spray powders should be used throughout the printing. Excessive plate sponging should be avoided (particularly the practice of laying a sponge on the extreme edges when the machine is in motion). Wiping may be a necessity in order to overcome catch-up, and pads of a more compact type are recommended.

F. D. Munday (PIRA) in the paper "Comparison of Synthetic and Plastics Paper Properties and Performance" said that, as a generalization, the essential features of a printing paper are that it should:

1. Give the required print quality without difficulty.
2. Be capable of being printed fast and without difficulty.
3. Look and feel acceptable to the reader.

Munday felt that new technologies may make current properties requirements with regard to (1) and (2) unnecessary, and reader re-education may change the requirements with regard to (3), but that "at present it is essential that the properties of any synthetic or plastics

42. Bulletin No. 80, "Litho Printing on Plastics," London, March 1972.

paper shall satisfy certain well-defined requirements that are met by appropriate conventional papers."[43] At the end of his presentation Munday listed certain conclusions:

1. Plastics and synthetic papers differ quite appreciably in their properties and performance.
2. If existing standards are applied, plastics and synthetic papers are superior to conventional papers in some respects but inferior in others.
3. Any strength advantage that has in the past been claimed for plastics paper needs very careful examination. Such claims tend to be overstated.
4. The relatively low stiffness of many plastics papers can be a serious disadvantage.
5. Using existing printing methods it is often more difficult to print plastics papers than conventional papers.
6. However, good print quality can be obtained, and advances in printing technology should make plastics papers easier to handle and print.
7. The consistency of plastics papers is no better than that of conventional papers.

Munday added that his presentation could not be regarded as comprehensive, and that only a few materials had been examined in detail.[44] There is no doubt that in the future, synthetic papers of different types, plastics papers of new grades, and products made from synthetic pulp will compete in certain cases for the same applications. But to quote Munday again, it must be stressed that "plastics are materials in their own right, and it is perhaps illogical that we should constantly have to think of them in terms of properties and performance that have been established for other materials—metals, wood, paper, etc."

PRODUCTION COSTS

In regard to the performance of the end products, it has to be taken into account that properties like folding endurance (for maps), water and chemical resistance (for manuals used in factories, children's books), weather resistance (for posters), etc., are properties that make synthetic papers superior to conventional papers for certain products. There are

43. In presentation at "Plastics and Paper" Conference, London, May 1973.

44. Among synthetic papers described in this book and not referred to in Munday's paper are YUPO FP, Phönix artplast, Arathene, Zexan, Spiax, Scott Plastic Paper, Syntosil, Artosil, Neobond, Pretex, Arval, Pryphane, and others.

other products, however, where the greater stiffness, incorporation of water-marks, etc., of conventional papers make them superior to synthetic papers. However, the main reason for the slow penetration of the market by synthetic papers has been the comparatively high price of synthetic papers compared with the price of conventional high-quality coated papers.[45]

During the introduction of synthetic papers the production costs were high because of the small capacity of the plants and their low utilization. It was estimated that in some cases production costs in a 12,000-tons/yr plant with a high production utilization were more than 10% lower than in the case of a 6,000-tons/yr plant. In a 15,000-tons/yr plant the production costs were approximately 7% lower than in a 10,000-tons/yr plant, or nearly 18% lower than in a 6,000-tons/yr plant. Synthetic papers aim—at least at present—at the speciality-applications market. It is a fragmented market, however, with a considerable total tonnage; it is therefore a challenge for this new industry. Some manufacturers have altered their marketing policy by reaching agreements with firms of the paper trade that have a detailed knowledge of the market. As a result of the fragmentation of the market, the introduction of new products, including the gradual overcoming of buyers' (and printers') resistance to using new materials, involves more time and cost. This is important since it influences the financial returns for manufacturers of these products.

The challenge to the synthetic paper industry is the penetration of the speciality markets of bigger tonnages, such as various sectors of publicity media (including manufacturers' catalogues and even insert pages in some journals), some sectors of the large-tonnage label market, and various stationery applications. At the same time, several manufacturers of synthetic papers have been very active in their R & D work aiming at the introduction of synthetic papers of second generation, which are expected to have superior properties and be more price-competitive.

The prices of conventional paper were uneconomically low during the 1970-71 recession and again during the 1975 recession, resulting in poor financial returns—and sometimes in actual losses—by some leading paper manufacturers. With greater demand for paper, paper prices have increased somewhat. But according to some authoritative sources these prices have not yet reached the proper remunerative level.

In the same way that plastics paper has been penetrating the low-priced packaging market, and some specialized plastics films have been achieving success in high-priced markets (e.g., capacitors and other

45. The price and quality competitiveness of "plastics paper" for wrapping and bag-making is described elsewhere in this book.

engineering applications) the synthetic paper industry will have to find for its products special markets in which the functional properties would justify a premium price, and the premium, because of technological development and bigger production runs, would be lower than at present.

The repercussions of the increased prices of the feedstocks on the synthetic polymer industry, along with the overall environmental and energy problems and their effect on the conventional paper and synthetic polymer industries, are referred to in the last chapter of this book.

4. Synthetic Pulp Processes and Feedstocks

Various processes have been developed and patented for producing synthetic pulp either from polymers or directly from monomers. In contrast to plastics-film-based synthetic papers and plastics papers produced on plastics processing equipment, synthetic pulp is made into paper on conventional papermaking machinery (sometimes with minor alterations to the machinery) regardless of whether the pulp is produced by fibrillation, ejection, or reactor fiber formation.

Although paper made exclusively from synthetic pulp can be produced, the emphasis, at least at present, is on the development of different types of paper combining synthetic pulp and conventional cellulose pulp. In this way paper is made that is superior in certain functions to paper made exclusively from cellulose or synthetic pulp.

Thus synthetic pulp, a nonwood fiber material, provides a further example of the growing cooperation and competition between the conventional pulp and the synthetic polymer industries. It is significant that several major developments in the field of synthetic pulp have been the result of joint activities of synthetic polymer and pulp/paper manufacturers. Prof. Bengt Rånby, of The Royal Institute of Technology (Stockholm), has referred to synthetic pulp as a new type of "potentially very inexpensive material" for the paper industry.[1]

The specific weight of polyolefins, at present the main feedstock for making synthetic pulp, is considerably lower than that of cellulose pulp (e.g., polypropylene with 0.91 and high-density polyethylene with about 0.96 g/cm^3), and this is an important factor.

Among various basic technologies to make synthetic pulp are:

1. The fibrillation method, which consists of orienting an extruded plastics film or tape to such an extent that fine fibrils are formed, or of

1. Fifth International Fundamental Research Symposium, Cambridge, England, September 1973.

beating cut pieces of oriented film into a fibrous structure similar to beaten wood pulp.

2. Flash-spinning methods, including melt spinning of polymers under high shear, and flash evaporation of an emulsion of polymer, solvent, and water.
3. Reactor fiber technology, which involves polymerization of an olefin to obtain a gel which is treated under high shear to produce nascent polyolefin fibers. This fibrous structure is subsequently refined into synthetic pulp.[2]

Several leading synthetic polymer and paper manufacturers have been developing synthetic pulp; the achieved stage varies from laboratory scale to semicommercial plant.

POTENTIAL APPLICATIONS

The functional properties of synthetic pulp make it an interesting additive to cellulosic pulp in the production of various specialty products, including wallpapers, embossed papers, tea-bag papers, water- and oil-resistant papers, dimensionally stable papers, wet-strength papers (e.g., filter papers), transparent papers, multilayer boards, electrical papers, etc., and in other applications such as diaper liners, battery separators, etc. An interesting feature of synthetic pulp is that its addition to the furnish of cellulosic pulp also has the advantage of improving the drainage rate in the papermaking process, and its use results in saving energy during beating and drying.

Thermoformable board made from recycled cellulosic materials and synthetic pulp offers interesting possibilities. Investigations are in progress for using synthetic pulp as an additive to groundwood pulp, since it is felt that in the furnish of groundwood and bleached softwood pulps it may be possible by addition of synthetic pulp to increase the groundwood pulp content and to decrease considerably the bleached softwood pulp content.

Synthetic pulp used as binder in nonwoven production is another potential application, since the bonding can be enhanced by thermomechanical methods.

Some manufacturers of specialty papers and nonwovens, interested in synthetic pulps, are continuing their R & D work on using synthetic pulps, while other manufacturers have adopted a "wait and see" policy.

2. Dr. William H. Hoge, "Synpulp Projects and Technologies" (Paperfilm Associates, Flemington, New Jersey, USA, 1973). R.E. Hichens, Shell International Chemical Co., "Prospects for the Application of Thermoplastics in the Communications Paper Industry" (Presentation at Ausplas, Australia, autumn 1973).

Some manufacturing firms, particularly in the engineering field outside the paper industry, also show interest in using synthetic pulp.

Although the impact of synthetic pulp on the market can hardly be expected to be felt before the early eighties, the potential of synthetic pulp looks promising. The development of synthetic pulps of second generation will result in a greater market penetration by synthetic pulp.

The very high tonnage consumption of conventional paper offers a potentially big market for synthetic pulp as an "additive" in a number of paper industry sectors. It is expected that a breakthrough in any sector would have a snowball effect on other applications of synthetic pulp.

On the question of the time factor related to the acceptance of synthetic pulp by the paper industry, it is worthwhile to remember the length of time it took to accept hardwood pulp. It is expected that in connection with the growing consumption of paper and board, the paper industry will be using nonwood fiber materials at an increased rate.

Synthetic pulp can be a valuable additive in the production of paper with special dielectric properties, whereas in other paper products its "addition" will result in better adhesion properties and decorative effects, and a higher rate of opacity and whiteness. The functional advantages which can be achieved in finished products by making them from conventional pulp *plus* synthetic pulp are expected to secure a market for synthetic pulp.

Two-ply paper from cellulosic and synthetic pulp, produced on a papermaking machine with a single head, with one layer mostly of cellulosic content and the other layer mostly of synthetic pulp, may become an interesting application of synthetic pulp. Tea-bag tissue containing synthetic pulp, and wallpapers with special properties due to synthetic pulp content (in both cases the synthetic pulp is combined with cellulosic pulp), are considered to be of potential interest.

A fundamental study on the finishing of synthetic pulp-containing paper in order to establish the relationships among smoothness, opacity, and compressibility, at varying temperature, pressure, and moisture, was undertaken by Crown Zellerbach.[3] In most cases blends of 40% synthetic pulp and 60% leaf bleached kraft pulp (LBKP) furnish were used. Full factorial experiments were run on paper containing 40% synthetic pulp, with the following variables:

Temperature (°C) 20, 50, 80
Pressure (kg/cm^2) 35, 140, 350
Moisture (%) 3, 5, 9

3. The bulk of the study was carried out at the University of Manchester Institute of Science and Technology (UMIST), Manchester, England.

The results of this study show that synthetic pulp-containing paper is finished more easily than conventional paper.[4] Paper containing 40% synthetic pulp is more compressible than the all-cellulose paper, and a relatively low nip pressure was needed to achieve the required finish. Even better paper smoothness was obtained using a resilient backing.

The calendering experience confirmed that papers containing synthetic pulps can be supercalendered to a smoother finish than 100% cellulosic paper while maintaining a higher bulk.

Moisture in paper containing synthetic pulp is associated with the woodpulp. Web moisture should be reduced below normal level before calendering in order to prevent excessive compaction and opacity loss of the woodpulp; the degree of moisture reduction should be proportional to the synthetic pulp content. Surface water on the web can be beneficial for developing smoothness without opacity reduction. This water was provided by steaming the web just before it reached the calender.

Commercial experience supports the findings of the Manchester study and shows that in most cases a decrease in calender pressure or the number of nips used is called for to maintain the same level of smoothness when using papers blended from synthetic pulp and woodpulp. Conversely, when maintaining the same calender conditions used for 100% cellulosic papers, the result is a smoother sheet.

POLYOLEFIN-BASED SYNTHETIC PULP

Crown Zellerbach

Crown Zellerbach Corporation (USA) developed synthetic wood pulp, called SWP, based on a direct or closely coupled process from monomer to fiber slurry. Ethylene and propylene were selected as the starting monomers, since they have been the lowest cost monomers.

In June 1971, Crown Zellerbach and Mitsui Petrochemical Industries (Japan) announced plans for a joint venture looking toward the manufacture of synthetic woodpulp for the Japanese domestic market. Mitsui Zellerbach K.K., of Tokyo, was formed, involving a 50:50 equity participation by both companies and an initial capitalization of 500 million Yen. The company combines the experience and technology accumulated by Crown Zellerbach in conventional paper and pulp activities with the experience and technology of Mitsui Petrochemical in polyolefin and petrochemical manufacturing. Utilizing Crown Zellerbach's SWP in-

4. Described by B.I. MacGowan, Crown Zellerbach Corp., in "Effect of Temperature and Pressure on Synthetic Pulp-Containing Papers," (Paper presented at the TAPPI International Synthetic Papers & Pulps Symposium, Washington D.C. March 31-April 2, 1976).

vention, Mitsui Petrochemical has been making extensive studies of the commercialized production process in the pilot plant for synthetic woodpulp at its Iwakuni-Ohtake Works. Crown Zellerbach has been making laboratory paper-machine runs in its Central Research and Development Department in Camas, which was extended during the development phase of the joint venture. In July 1972 Crown Zellerbach and Farbwerke Hoechst of Frankfurt (Fed Rep Germany) signed an agreement "looking forward to the establishment of a joint venture for the production and marketing of synthetic woodpulp (SWP) in Europe," with the aim of serving markets in the enlarged European Economic Community and in Finland. During the period of market development in Europe, in order to determine the commercial feasibility of manufacturing SWP there, development quantities of SWP have been supplied by the semicommercial plant in Japan.

SWP Properties

Crown Zellerbach feels strongly that papers made on conventional paper machines from cellulose pulp with an SWP content result in a broad application potential that cannot be achieved by pure cellulose pulp or pure synthetic fiber products. Some advantages of SWP are:

1. SWP fibers are water dispersible and thermoplastic, and their shape and size distribution is somewhat similar to conventional woodpulp. The developed process enables to some degree the control of fiber dimensions and geometry; it is therefore possible to duplicate the fiber-size distribution corresponding to a typical hardwood or softwood pulp.
2. The high surface area of SWP (high compared with that of conventional synthetic fibers) is an important feature of this new product.
3. SWP melts at temperatures of about 130° to 135°C.
4. Heat bonding can increase the web strength of SWP sheets. In addition, starches, melamine resins, and synthetic polymers can be used to improve the tensile strength of paper made from cellulose pulp with an SWP content.
5. The SWP content is chosen in accordance with the required properties of the end product, whereby the variations of the SWP content do not affect the dispersibility of the SWP fibers.
6. The addition of SWP fibers to conventional pulp has the effect of higher brightness and opacity.
7. The porosity of the end product increases with the SWP content, an important factor in filtration applications.
8. Because of the inertness of polyethylene to moisture, SWP furnishes provide for dimensional stability during changes in humidity.

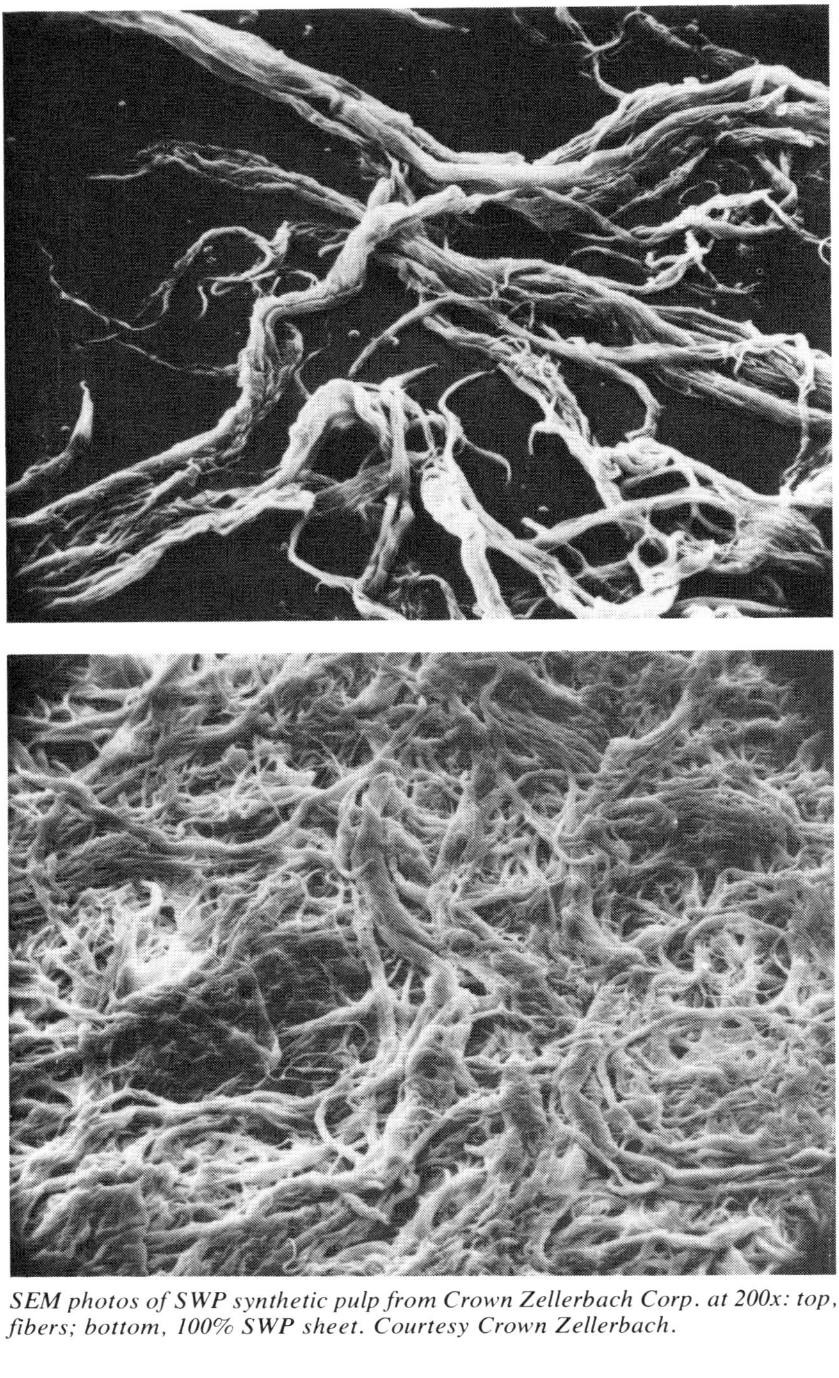

SEM photos of SWP synthetic pulp from Crown Zellerbach Corp. at 200x: top, fibers; bottom, 100% SWP sheet. Courtesy Crown Zellerbach.

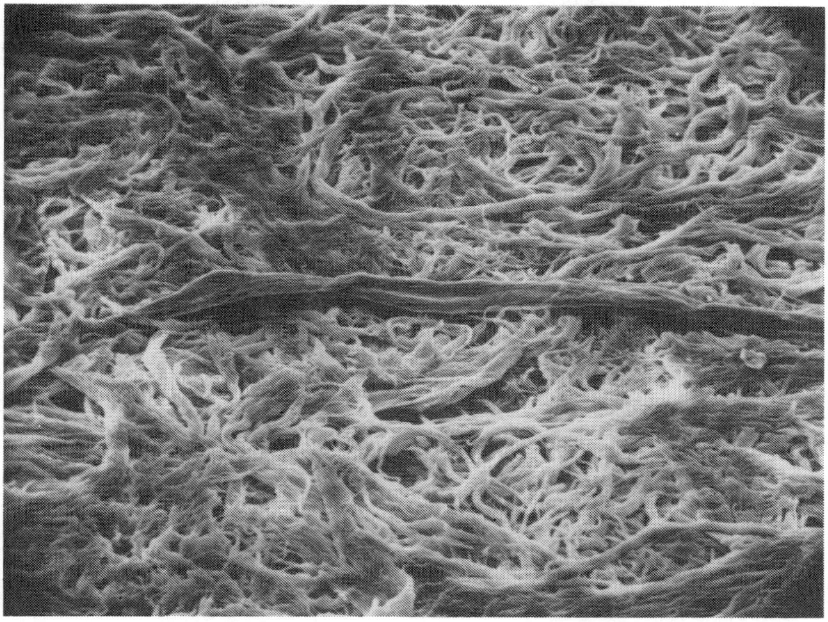

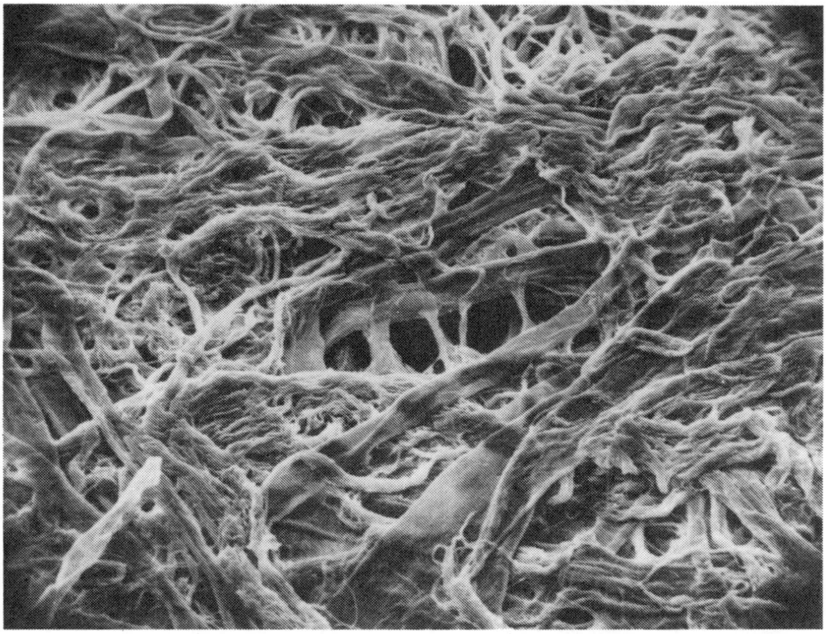

SEM photos of blended sheet containing: top, 95% SWP; bottom, 50% SWP, 20% softwood pulp, and 30% hardwood pulp at 200x. Courtesy Crown Zellerbach.

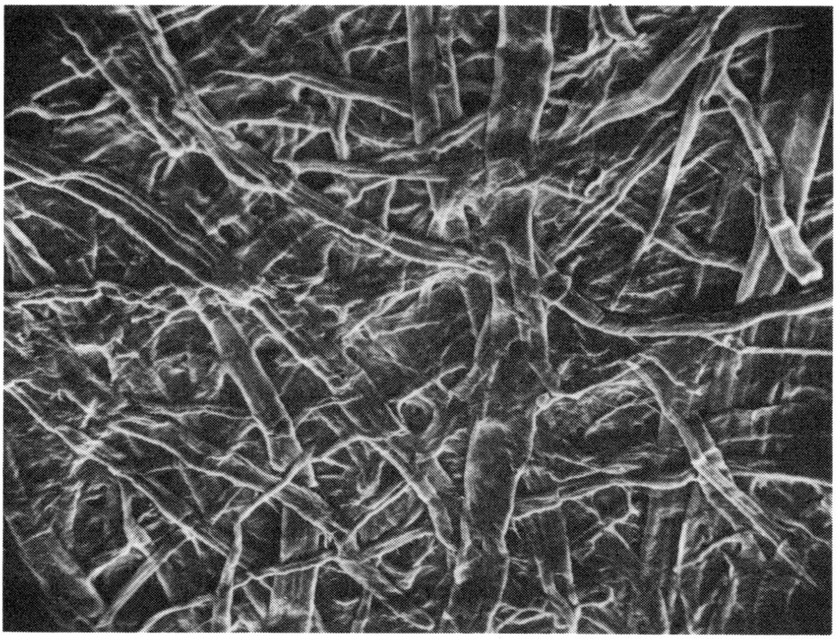

SEM photo of a 100% cellulose sheet at 200 x. Addition of SWP gives higher brightness, opacity, and porosity. Courtesy Crown Zellerbach.

9. Papers with an SWP content have a much higher bulk than all-cellulosic pulp paper, and SWP blends can therefore be calendered to a very smooth finish and still maintain a high bulk (the density should not exceed 0.7 g/cc, since it is at this point that blackening starts and opacity decreases strongly).
10. Paper with an SWP content shows improved optical properties, which are important in printing papers where minimum show-through is desired. Crown Zellerbach has carried out a number of machine trials on lightweight opaque grades. These have been printed by rotogravure, offset, and letterpress with, according to Crown Zellerbach, excellent results. The thermoplastic nature of SWP can be used to advantage in a so-called "thermal watermark." All that is needed to achieve thermal watermarking or thermal printing is a heated die.
11. The water resistance of paper with a SWP content is expected to result in the penetration by this paper of a number of speciality-product markets. It is possible to make water-resistant papers having a wide range of porosity degrees. Similarly, the opacity of this paper can be varied from totally opaque to almost transparent. When

Table 4.1 *Comparison of Electrical Papers*

Material	*Basis weight* (g/m^2)	*Caliper (micron)*	*Density* (g/cm^3)	*Breakdown strength*[a] *(kV/mm)*	*Dissipation factor*[b]
100% SWP	60	79	0.76	59	0.0013
Coil insulation	59	78	0.76	10	0.0137
PE	–	–	0.92	50[c]	0.0003[c]

[a] Dielectric breakdown strength. Mean of 10 determinations at 60 H_Z with 25-mm electrodes loaded to 0.5 kg/cm^2.

[b] Dielectric dissipation factor (tangent of the angle of dielectric losses). Mean of 5 determinations made at 10^3 H_Z.

[c] Literature values. Dissipation factor applies to 10^3 through 10^9 H_Z.

Source: J. S. Barton and T. Matsuda, "Synthetic Wood Pulp – a New Raw Material for Papermaking," *Man-Made Polymers in Papermaking: Proceedings of the IUPAC-EUCEPA Symposium* (Helsinki: Finnish Pulp and Paper Research Institute, 1973), p. 149.

the SWP content is above approximately 60%, the sheets become readily heat-sealable, indicating their potential use in packaging overwraps, heat-sealing pouch stock, etc.

12. A heavily calendered 100% SWP web with sufficient temperature treatment produces a tough film which offers a potential alternate approach to making packaging films (other than by the conventional extrusion process). Crown Zellerbach expects that SWP will also find application in the nonwoven area as a bonding agent when blended with other synthetic and natural fibers.

Table 4.1 compares the electrical properties of SWP with those of insulating paper and polyethylene film and indicates the potential use of SWP paper in the field of insulating applications.

Potential Applications

Crown Zellerbach lists the following potential applications of SWP products:

- ☐ Water-resistant packaging paper.
- ☐ Water-resistant liner.
- ☐ Water-resistant printing and writing paper: maps, signs, tags, labels, notebook paper (surveyors, etc.), envelopes, others.
- ☐ Dimensionally stable papers: recording papers, wallpapers, IBM cards, base for carbonizing.

- ☐ Wet-strength paper: filter paper, membranes for dialysis, reverse osmosis, hot-house paper; medical tape, nursery paper.
- ☐ Decorative papers: playing cards, artificial parchment, tape, Shoji paper.
- ☐ Others: paper plates (and other moulded articles), paper cups, release paper, diaper liners, photographic paper, children's book paper, battery separators, electrical papers, art paper, synthetic leather base paper, heat shrink paper, tea-bag paper, protective papers, oil- and grease-resistant paper.

Table 4.2 *The Effect of SWP Addition on Properties of Machine-Made Papers*

	SWP level (%)				
Properties measured	*0%*	*7.5%*	*20%*	*40%*	*90%*
Physical properties					
Basis weight (g/m^2)	59.4	52.7	50.1	61.5	60.2
Caliper (mm)	0.089	0.090	0.135	0.119	0.135
Density (g/cc)	0.659	0.591	0.636	0.514	0.444
Sugar dye (sec)	–	–	–	–	–
Optical properties					
Opacity (%)	64.9	69.3	83.6	84.7	85.7
Brightness (Elrepho #8)	77.0	78.0	83.7	88.9	94.0
Mechanical properties					
Tensile (kg/15 mm)					
MD	7.10	6.35	3.27	4.42	2.7
TD	3.70	3.51	1.63	2.20	1.8
Stretch (%)					
MD	2.2	1.9	1.3	1.5	2.1
TD	4.1	4.1	2.8	3.1	3.5
TEA (g · cm/cm^2)					
MD	73.0	46.0	31.0	27.0	11.0
TD	80.0	65.0	34.0	31.0	12.0
Tear (g/sheet)					
MD	35	28	17	27	17
TD	41	34	19	30	19
Scott internal bond	206	189	96	92	182

Note: Machine-made papers were composed of SWP-LBKP as follows:

SWP	0	7.5	20	40	90
LBKP	100	92.5	80	60	10

Source: Barton and Matsuda, "Synthetic Wood Pulp – a New Raw Material," *Man-Made Polymers in Papermaking: Proceedings of the IUPAC-EUCEPA Symposium*, p. 145.

Table 4.2 shows the effect of SWP addition on properties of machine-made papers, and Table 4.3 shows the wet-end additives used to produce lightweight offset bond and the properties of base paper with and without SWP.

Table 4.3 *Wet-End Additives Used to Produce Lightweight Offset Bond and Properties of Base Paper With and Without SWP*

Properties measured	*Control*	*50% SWP*
Basis weight (g/m^2)	45.1	40.8
Caliper (mm)	0.073	0.089
Apparent density (g/cc)	0.611	0.457
Tear (g/sheet)		
MD	13	17
TD	16	20
Fold		
MD	10	18
TD	4	5
Tensile (kg/cm)		
MD	1.86	1.82
TD	0.91	0.85
Stretch (%)		
MD	1.10	1.32
TD	1.73	2.48
Tensile energy absorption (g-cm/cm^2)		
MD	11.0	18.0
TD	11.0	15.0
Sheffield smoothness (cc/min)		
Wire side	296	329
Top side	254	302
Gurley densometer (sec)	300	150
Sugar dye size (sec)	6	16
Tappi opacity (%)	77.5	76.2
Brightness (Elrepho #8)	78.1	85.6
Scott internal bond (kg-cm/cm^2)	351	279
Fiber furnish (%)		
SWP	0	50
NBKP	40	30
LBKP	60	20
Wet-End additives		
Clay	Present	0
Rosin (%)	1	0.5
Cationic starch (%)	0.5	0.5
Melamine resin (%)	1	1
Alum to pH	4.8	4.8

Source: Barton and Matsuda, "Synthetic Wood Pulp—a New Raw Material," *Man-Made Polymers in Papermaking: Proceedings of the IUPAC-EUCEPA Symposium*, p. 146.

Table 4.4 *Typical Physical Properties of Various SWP Pulp Grades*

	Grade					
	E-400	*E-600*	*E-620*	*E-790*	*R-830*	*R-990*
Drainage factor (sec/g)	2	2	6	<1	10	<1
Classified fiber length (mm)	0.9	1.3	1.3	1.6	2.0	2.5

Source: B. I. MacGowan, "Effect of Temperature and Pressure on Synthetic Pulp-Containing Papers" (Presentation at the TAPPI International Symposium on Synthetic Papers & Pulps, Washington, D.C., March 31-April 2, 1976).

Notes: E = commercially available, R = development stage (early 1976). Grades E-790 or its longer counterpart 990 are preferred in the nonwoven production, whereas R-830 is preferred where a dense, dimensionally stable, strong, and coatable surface is desired, e.g., in tags and labels.

Crown Zellerbach has continued the development of its synthetic pulp SWP; Table 4.4 shows the physical properties of the company's four polyethylene grades available for commercial purposes and its two grades at development stage in early 1976.

Among patents issued to Crown Zellerbach is "Process of Forming Polyolefin Fibres,"[5] which refers to the previous Crown Zellerbach patent "Process for Polymerising Olefins to Form Fibres"[6] according to which polyolefin fibers are formed directly from olefin monomers during polymerization. The more recent patent describes a process of making polyolefin fibers of very high molecular weight by first forming a fibrous gel that does not contain discrete (free-floating) fibers, and then subjecting the gel to a relatively high shear stress, causing discrete fibers to appear.

The application of synthetic pulp as a reinforcement for plastics films has been patented by Crown Zellerbach, and this patent describes various methods that can be employed to incorporate pulplike synthetic fibers into the thermoplastics film with the aim of producing heavy-duty bags and packages.[7]

Solvay

Solvay & Cie. (Central Research Office in Brussels) has developed a process for making synthetic pulp from microfine high-density

5. British Patent 1,350,487, Complete Specification published April 18, 1974. See Wolpert, *Plastics and the Paper Industry*, pp. 31-35.

6. British Patent 1,287,917, Complete Specification published September 6, 1972.

7. British Patent 1,360,115, Complete Specification published July 17, 1974. See Wolpert, *Plastics and the Paper Industry*, pp. 57-58.

polyethylene fibers for wet processing. Solvay feels that synthetic pulp will find application, mainly as an additive in the production of various types of paper products. In the case of printing and writing paper the addition of synthetic pulp to cellulose pulp adds to opacity and brightness of the end product, and in this case synthetic pulp can replace fillers and/or coatings (by adding about 20% of synthetic pulp content to the furnish). The additional advantage of synthetic pulp content is the decreased specific weight of the end product. In the case of wallpaper, a large-scale potential application, synthetic pulp addition results in an increased dimensional stability, and strippable wallpaper coverings can be made from a 100% synthetic pulp furnish.

Multilayer boards consisting of two-plies of bleached kraft or sulphite with a center layer made for PE pulp is an interesting application possibility due to the functional properties of this new type of board.

Up to the end of 1973, Solvay spent more than BF300 million on the development of synthetic pulp. According to the information supplied by Solvay in early 1974, the company had signed agreements with about 20 paper manufacturers, which were evaluating trial quantities of Solvay pulp and reporting to Solvay on the special quality requirements of this pulp and on the approximate size of the market (depending on the price of synthetic pulp) for individual applications. In this way cooperation was established with these papermaking firms, some of which are manufacturers of speciality types of paper. In many cases the agreement precludes the release of the name of the firm that has entered into this arrangement with Solvay, and in some cases not even the country of the papermaker is allowed to be announced.

Among the firms that signed agreements with Solvay were:

- ☐ Three Finnish companies: including Kymmene AB, Tampella AB.
- ☐ Two Belgian companies: including Genval (for its wallcovering material plant).
- ☐ One Italian company: Cartiere Burgo, SpA.
- ☐ Two French companies: including one manufacturer of diazopapers (negotiations with two other French companies had reached an advanced stage by the end of 1973).
- ☐ Four British companies.
- ☐ One West German company.
- ☐ One Swedish company.

A special agreement was concluded with a Japanese chemical company for evaluation of the Japanese market, with the possibility of this manufacturer producing this pulp under the Solvay license. By the end of 1973 talks with a U.S. tea bag manufacturer were in progress.

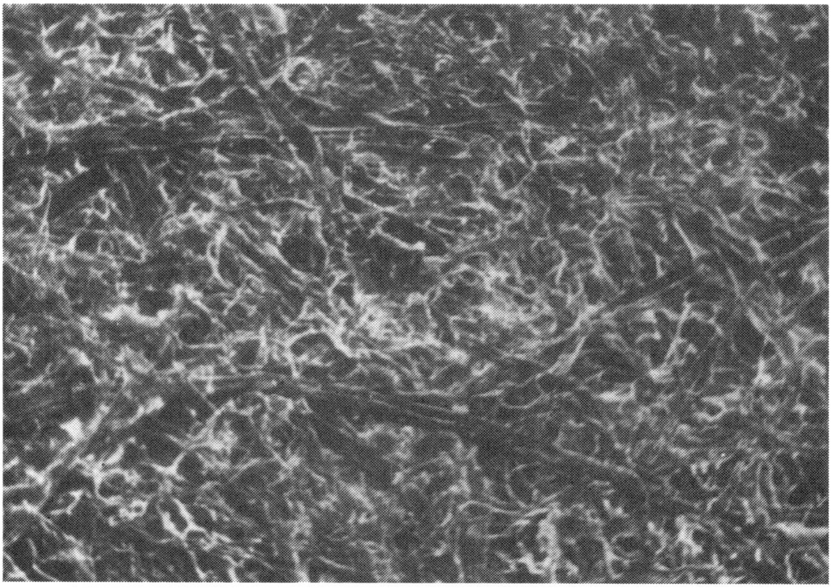

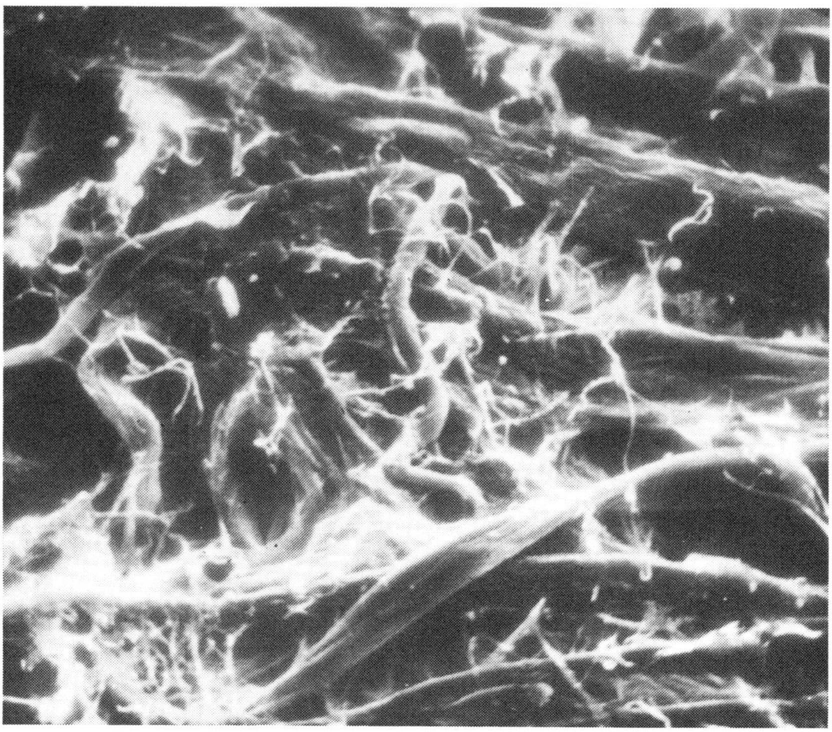

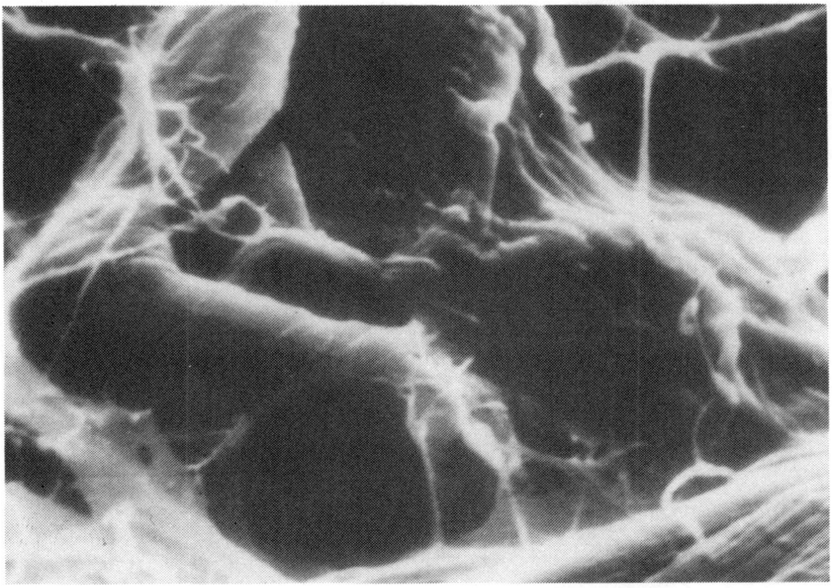

SEM photo of paper made with 30% synthetic pulp from Solvay & Cie. of Belgium and 70% cellulose pulp at: top of facing page, 100x; bottom of facing page, 1000x; above, 5000x. Courtesy Solvay.

Production Method and Properties

Solvay uses ethylene as the main raw materials, and the conversion into synthetic pulp is performed in a continuous process, consisting of the following stages:

1. Polymerization at low pressure using a supported catalyst with a very high yield. The purification of the polymer from catalyst residues (applied in conventional PE production) is not required, thus resulting in low manufacturing cost.
2. Conversion of the polymer into pulp. This stage of the process, which cannot be separated from the first stage, involves heat and pressure release, with the addition of water and of some minor quantities of other additives, in order to obtain an aqueous slurry of fibrous substance (the diluent and unreacted monomer are separated and recycled to stage one).

The produced slurry can be processed in an integrated paper plant, or it can be dewatered and the PE pulp conditioned in the same way as conventional woodpulp. The same process with some minor modifications can be applied for using propylene or ethylene together with

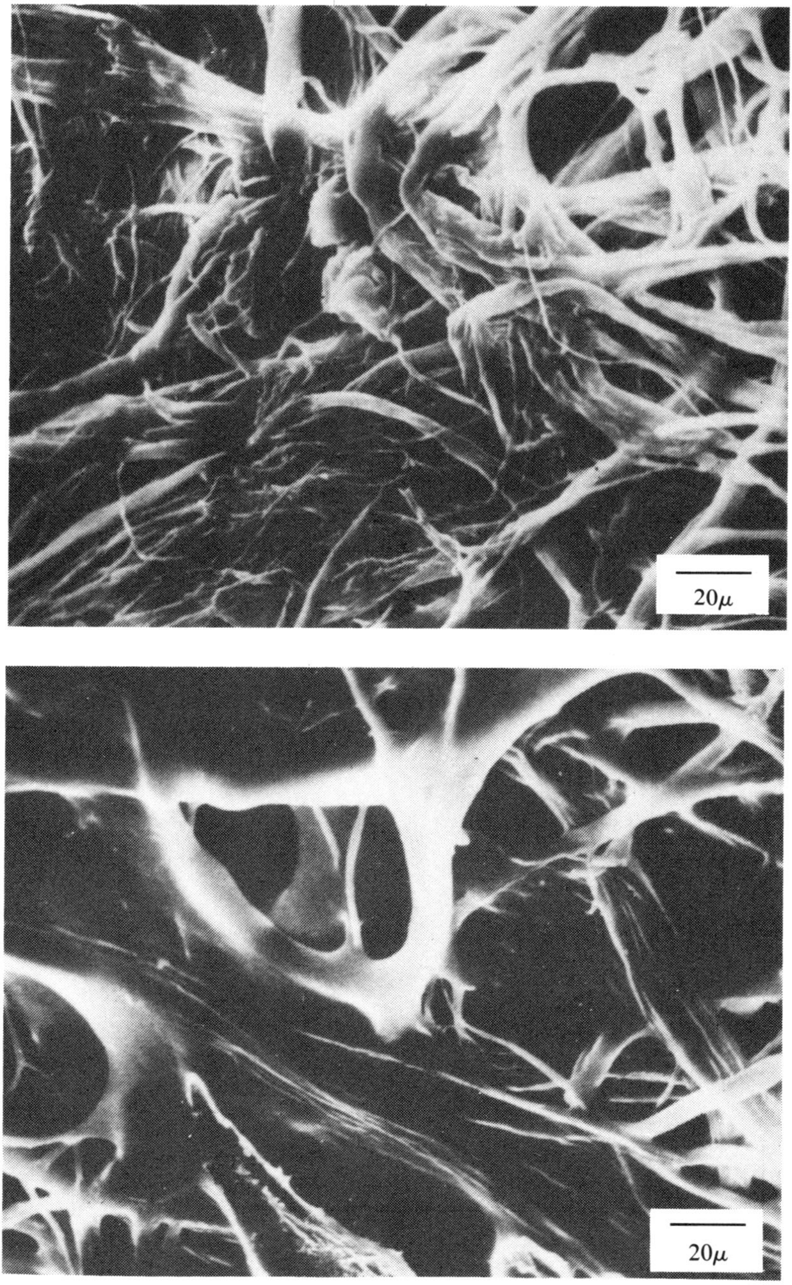

SEM photos of sheet made of 100% Solvay synthetic pulp; top, untreated sheet; bottom, after thermal consolidated treatment. Courtesy Solvay.

propylene as feedstock, with the properties of the pulp being suitable for particular applications.

The PE pulp consists of extremely fine fibrils (thickness about 1 micron or less; the length can be controlled within certain limits) and has a very high surface area (30 m^2/g or even higher, as against less than 0.2 m^2/g in the case of most synthetic fibers obtained by conventional spinning techniques). The tenacity is in the range of 1.5 to 2 g/decitex, which corresponds to a breaking length of 15,000 to 20,000 m.

Polyethylene is known to be water repellent, which is a serious drawback as far as wet processing is concerned; the nonpolar characteristics of polyethylene have therefore to be modified in order to make it wettable. The Solvay process is claimed to achieve such a modification at a very low additional cost.

The specific gravity of high-density polyethylene (0.96) compares favorably with that of cellulose fibers (1.5), which means that for the same thickness of paper, the yield expressed in square meters per unit of weight is about 35% higher for PE pulp than it is for woodpulp. In the case of polypropylene, which has a specific gravity of 0.91, this advantage is even greater.

PE pulp can be beaten with conventional refiners. However, because polyethylene cannot be hydrated, the effect of refining for synthetic pulp is purely physicial—that is, it results in separation of individual fibers and in shortening of fibers. Low beating gives long fibers with average dispersion and average look-through, whereas hard beating gives short fibers with optimum dispersion and evenness.

PE pulp can be mixed with woodpulp in various proportions either before or after refining, and unlike conventional synthetic fibers, it can be easily processed on unmodified papermaking machines (Table 4.5 shows the experimental machine runs).

Once the sheet has been formed, it is necessary to consolidate it by binding the fibers together. The chemical bonding of cellulose fibers is quite effective and requires only the presence of water. Unmodified polyethylene, on the other hand, cannot be bonded in this way. In fact, the chemical inertness and nonreactive nature of polyolefins make it difficult to bind polyolefin fibers with one another or with cellulose fibers. Papers based on PE pulp can be consolidated in two ways:

1. By putting to advantage the thermoplastic nature of polyethylene, that is, by partial and local fusion of the fibers at their points of contact.
2. By a chemical modification of the polyethylene base, that is, by introducing polar groups which are apt to react between themselves and with the polar groups of cellulose fibers.

Table 4.5 *Experimental Machine Runs (table width 0.5 m)*

	Feedstock				
	100% PE pulp 17° SR		*40% PE pulp, 60% bleached kraft 32° SR*		*30% PE pulp, 70% bleached kraft 28° SR*
Speed (m/min)	32	45	32	32	45
Consistency (%)	0.26	0.30	0.24	0.19	0.22
Weight (g/m^2)	55	60	65	45	60
Draining rate	very high	very high	high	high	high
Wet strength	low[1]	good	good	good	good

1. Insufficient for weights below 45 g/m^2.

Source: Information supplied by Solvay (quoted in Wolpert, *Synthetic and Plastics Paper,* Vol. 3, 1972, p. 464).

Table 4.6 *Typical Properties of Papers Made of 100% PE Pulp*

Properties	*Type 1*	*Type 2*
Weight (g/m^2)	59	78
Thickness (mm)	0.123	0.165
Specific volume (cm^3/g)	2.10	2.15
Breaking length (m)		
MD	4,400	4,600
TD	3,600	3,700
Elongation (%)		
MD	22	26
TD	17	20
Tear index		
MD	162	115
TD	184	131
Folding endurance		
MD	3,210	>4,000
TD	2,620	>4,000
Brightness (%)	86.6	88.5
Opacity (%)	83.8	89.8

Source: Information supplied by Solvay (quoted in Wolpert, *Synthetic and Plastics Paper,* Vol. 3, 1972, p. 465).

Table 4.7 *Mixtures of PE Pulp with Woodpulp*

	70% bleached kraft, 30% PE pulp		*70% groundwood, 30% PE pulp**		*35% unbleached sulphite, 35% groundwood, 30% PE pulp†*
Heat consolidation	*No*	*Yes*	*No*	*Yes*	*No*
Weight (g/m^2)	62	63	60	61	96
Thickness (mm)	0.127	0.125	0.125	0.125	0.170
Specific volume (cm^3/g)	2.05	2.0	2.10	2.05	1.77
Breaking length (m)					
MD	4,250	5,370	1,700	2,430	4,420
TD	1,720	2,130			2,160
Elongation (%)					
MD	1.7	2.1	0.5	1.0	1.8
TD	5.1	5.7			3.3
Tear index					
MD	123	125	48	43	77
TD	151	147			85
Brightness (%)	88	83	74	68	–
Opacity (%)	88	74	96	88	97
COBB test (g/m^2/60s)					
Glazed side					11
Dull side					11
Dimensional stability					0.9‡

* Handsheets.
† Plus 5% talc plus 1.5% sizing.
‡ Percent dimensional change after immersion in water for eight minutes.

Source: Information supplied by Solvay (quoted in Wolpert, *Synthetic and Plastics Paper,* Vol. 3, 1972, p. 466).

Typical properties of a 100% synthetic paper based on PE pulp are listed in Table 4.6. This paper has been consolidated by heat (i.e., local fusion of the fibers). The properties in this table refer to measurements in a dry condition, but since PE is unaffected by moisture, the same characteristics are valid also in wet conditions; moisture content does not influence the dimensional characteristics of this paper.

Optical properties (opacity and brightness) are among the main advantages of PE pulp paper and are obtained without any bleaching or fillers. The opacity is related to the fineness of the fibers or to their surface area in the same way that opacity is related to the particle size of fillers.

The properties of paper made from mixtures of PE pulp and woodpulp depend on the relative proportions, the kind of woodpulp, and the kind of consolidation. Table 4.7 shows the properties of papers made from various mixtures, either heat consolidated or not heat consolidated, including mixtures of PE pulp with bleached kraft, PE pulp with groundwood, and PE pulp with sulphite and groundwood.

Except in the case of groundwood, where the addition of PE pulp improves the mechanical properties, strength properties are impaired by PE pulp addition. However, more recent research seems to indicate that chemically modified PE pulp can give much greater strength properties. The main improvements to be found, as already mentioned, are in optical properties, dimensional stability under wet conditions, and heat sealability.

Table 4.8 contains some data on multilayer boards made of one layer of PE pulp between two layers of woodpulp and shows the possibility of obtaining waterproof board that has both high dimensional stability and high cohesion.

Papers made from PE pulp can be reprocessed as long as the fibrous nature has not been altered too much. Papers made from more than 50% PE pulp, which have been consolidated, are not easily reprocessed, but papers containing lower percentages of PE pulp can be reprocessed even if they have been moderately heat treated.

Solvay calls its synthetic pulp Pulpex, and although most of the development work has been done with HD PE (Pulpex E), the company also offers Pulpex P, based on polypropylene, saying that each type offers particular advantages and original characteristics, but that both types have numerous features in common. Table 4.9 lists the corresponding values of properties for the two polyolefins.

Improved Wettability

Solvay's latest development work has been directed towards improved wettability of synthetic pulp.[8] Because of the hydrophobic character of polyolefin fibers their use in papermaking requires a wettability sufficient to prepare an aqueous suspension. Fibers of 3- to 10-mm length produced by flash spinning are collected in a flasher containing water and wetting additives. The aqueous suspension passes through a refiner and is then concentrated on a thickener which is in fact a very simplified paper machine. This prerefined pulp, containing roughly 60% water, is packed and dispatched to the paper mills, where it is either added to a prerefined cellulose *or* added to the cellulose pulp and refined before the sheet-forming operation takes place. Improving the wettability of synthetic pulp in this way has a number of drawbacks, including, at

8. See Dr. J.P. Bex, "New Developments in Synthetic Polyolefin Pulps" (Presentation at the TAPPI International Symposium on Synthetic Papers & Pulps, Washington D.C., March 31-April 2, 1976).

Table 4.8 *Multilayer Boards*

Structure	*First layer (g/m²)*	*Bleached kraft 60*				*Bleached sulphite 60*
	Second layer (g/m²)	*PE pulp*				*60% PE pulp 40% bleached sulphite*
		60	*60*	*40*	*30*	*60*
	Third layer (g/m²)	*Groundwood 60*				
Cohesion (Scott)		>250	>250	>250	>250	>250
Water vapor transmission $g/m^2 \cdot 24$ hr (38°C, 90% RH)		6.4	5.5	15	47	34

Source: Information supplied by Solvay (quoted in Wolpert, *Synthetic and Plastics Paper,* Vol. 3, 1972, p. 467).

Table 4.9 *Values for HD PE and PP*

Property	*Unit*	*HD PE*	*PP*
Specific gravity	g/cm^3	0.965	0.900
Melting point	°C	132	165
Water absorption	%	0	0
Maximum tensile strength	kg/cm^2	360	330
Tensile strength at yield	kg/cm^2	240	330
Elongation at break	% at 50 mm/min	>170	>200
Elastic modulus	kg/cm^2	21,800	12,000
Dielectric constant		2.4-3.5	2.2-2.3
Loss factor	tg δ	$<5.10^{-4}$	3.10^{-4}
Dielectric strength	kV/cm	4.10^2	3.10^2
Transverse resistivity	Ω cm	10^{17}	10^{16}

Source: Solvay & Co. "Characteristics and Use of the Synthetic Paper Pulp Pulpex," publication NP 28C-B-2-376, 1976.

least to some extent, the concealment by the wetting agents of some important properties of the pulp. In addition, the drying of synthetic pulp containing wetting agents is not advisable, since redispersion may become a problem.

Solvay has developed a way to overcome the above shortcomings, using a small amount of cellulose as a wetting additive (e.g., 2% of the synthetic pulp weight) in a classical pulper, which results in a complete wettability of the fibers. The cellulose-wetting process results in synthetic-pulp-containing papers with properties at least as good as those obtained by the chemical-wetting process. In fact the breaking length of the paper made from cellulose-wetted pulp improves the breaking length of the paper. The new method enables the collection of dry fibers in a special collecting device and their supply in this form to the papermakers, which results in cost reduction in the areas of production and transportation.

When adding a small percentage of cellulose to synthetic pulp, and then adding this modified synthetic pulp to cellulose pulp, a stability difference between the two components remains which can be utilized for the production of twin-ply papers on a single-head machine. This two-ply product has interesting applications, since the ply which mainly contains synthetic fibers is heat-sealable. In addition, this production method has advantages over the method of first producing a paper sheet and then coating it with plastics to achieve heat-sealability.

In most cases, however, the stability difference is a disadvantage, but it can be rectified by a further refining of the mixture of cellulosic pulp and synthetic pulp.

Polyethylene or Polypropylene

At the 1972 IUPAC-EUCEPA Conference in Helsinki,[9] in the discussion that followed the presentation entitled "Synthetic Pulp for Wet Processing Made of Microfine High Density Polyethylene Fibres" by Michel Osterrieth (Solvay & Cie) and "Synthetic Wood Pulp—A new Raw Material for Papermaking" by Dr. J.S. Barton (Crown Zellerbach Corporation) and T. Matsuda (Mitsui Petrochemical Industries), the question of polyethylene copolymers and polypropylene as raw material was raised.

Dr. Barton commented that in his presentation he had referred only to PE, and added that Crown Zellerbach has done some work on PP too,

9. *Man-made Polymers in Papermaking: Proceedings of the IUPAC-EUCEPA Symposium, Helsinki, June, 1972*. For Osterrieth, see pp. 121-31; for Barton and Matsuda, see pp. 133-50; for the discussion of the two presentations, see pp. 151-56.

and that in cases where higher heat resistance was required PP would be used.

M. Osterrieth commented that PP was less stable to sunlight, and, therefore, the use of PP in some applications would be inappropriate. He felt that ultimately the best performance should be reached by a mixture of PE and PP (Solvay has done work on PP and mixtures too).

Bonding of synthetic pulp webs is an important factor, and the tensile strength of the fiber must be chosen correspondingly. Further R & D work on thermoconsolidation of the process has been in progress.

During the discussion at the Helsinki Conference Dr. Barton stressed the point that in the SWP process a conversion to a hydrophilic material was achieved, and that in a beater filled with pulp mixed with synthetic material it was not possible to see the difference of the materials, since they were both hydrophilic materials.

From the point of view of price, the use of propylene as a raw material for making synthetic pulp has been becoming more advantageous than the use of ethylene, and this trend is expected to continue. PP also has a lower specific weight than HD PE, which is a further advantage. With the development of the synthetic pulp market, paper manufacturers will have the choice of PE, PE/PP, or PP synthetic pulp, depending on the end product.

Economy of Size

On the question of price for synthetic pulp, M. Osterrieth reported that the investment for a plant producing 100,000 tons/yr of PE pulp was approximately the same as for a plant manufacturing the same amount of HD PE plastics material. The production costs, including the consumption of raw materials, labor, etc., were approximately the same for PE pulp and PE plastics. Only a few minor modifications were required to use the same process for propylene instead of ethylene.

Dr. Barton also reported that in the company's relatively small plant in Japan the price of synthetic pulp was about twice the price of bleached long fiber kraft pulp in Japan; he added that in a plant the size to which Osterrieth had referred the picture would change, with the economics of scale reducing the production cost.[10]

---

10. In "Synpulp Projects and Technologies" (see footnote 2), Dr. William H. Hoge says that the joint Mitsui-Zellerbach plant supplies two grades of SWP, E 400 and E 600, which differ in fiber length. This report includes the condensed translation of the Crown Zellerbach German Patent Application covering methods to make water-dispersable polyolefin fibers and says that the application was filed on February 23, 1972. The publication date of the application was September 14, 1972 (German Patent Specification 2,208,555). According to Dr. Hoge's report, this application corresponds to the U.S. Patent Application 120680 filed on March 3, 1971.

NEWCOMERS

Among other manufacturers who have developed their own processes of making synthetic pulp and have recently established semi-commercial plants are ANIC of San Donato Milanese (Italy) and Montedison of Milano (Italy). Other European, U.S., and Japanese manufacturers have been pursuing R & D work in this area on a laboratory scale.

ANIC

ANIC has installed a semi-commercial plant at Ragusa, Sicily, with a process that enables the use of HD PE, PP, HD PE plus PP, and HD PE plus PS, as well as of PVC, ACN (acrylonitriles), and polyamide-6 for making synthetic pulp.[11] The results of tests carried out on paper containing 66% cellulosic pulp and 34% polyamide-6 synthetic pulp proved to be satisfactory, but the high cost of polyamide compared with that of HD PE seems to rule it out as an economic proposition.

The company's synthetic pulp based on HD PE is called PEAD. The microfibers which constitute the synthetic pulp appear in the form of plexifilamentary aggregates with high specific surface values (30-50 m^2/g). The fibrous aggregates are prevalently constituted by fibers between 2 and 4 microns in diameter. However, more compact groups of filaments, whose maximum diameter does not exceed 10 microns, are to be observed. The average length of the single fibers is approximately 2-3 mm. The length of these fibers can be varied during production.[12]

Synthetic pulp can be "beaten" alone or in mixture with cellulose. When it is to be beaten alone, it requires small quantities (up to 5% based on the synthetic pulp weight) of cellulose, or approximately 1% of carboxyl methyl cellulose, an additive used in the paper industry. This beating process (different from that used for cellulose), leads to the splitting of the fibrous bands and a shortening of the single-fiber length.

ANIC is engaged in development work to produce synthetic pulps containing hydrophilic groups. These new synthetic pulps, also called "reactive" pulps, should show a considerable affinity to water. As a result of these hydrophilic groups, they interact with the cellulose, resulting in improvement of mechanical properties of the paper made from a blend of cellulosic pulp and this "reactive" pulp. ANIC feels that it has succeeded—using the same production process as the company uses for traditional synthetic pulp—in producing a modified synthetic pulp which has a better cohesion with cellulosic pulp. The advantage of

11. See Wolpert, *Plastics and the Paper Industry*, pp. 40-44.

12. G. Serboli and V. Foti, ANIC, "Synthetic Pulp in the Paper Industry" (Presentation at the TAPPI International Synthetic Papers & Pulps Symposium, Washington D.C., March 31-April 2, 1976).

Table 4.10 *Comparison of Physical Properties of Handsheets Containing Synthetic Pulp*

Property	*Unit*	*Synthetic Pulp*		
		(0%)	*(34% HD PE)*	*(34% modified HD PE)*
Basis weight	g/m^2	64	61	65
Thickness	mm	0.100	0.115	0.123
Apparent specific volume	cc/g	1.55	1.88	1.90
Breaking length	m	5,400	2,400	3,600
Stretch	%	3.4	3.2	4.5
Tear factor	m^2	114	100	98
Opacity	%	77.2	90	90
Brightness	%	80	88	87

the hydrophilic groups is that they make possible "excellent wetting and water dispersibility of modified synthetic pulp" without the use of additives. Table 4.10 compares the physical properties of paper containing 100% cellulosic pulp, 66% cellulosic pulp and 34% synthetic pulp, and 66% cellulosic pulp and 34% modified HD PE synthetic pulp. This table shows that synthetic pulp containing hydrophilic groups results in an increase of the breaking length of the paper, while, on the whole, the other properties remain constant.

The increase of the breaking length of the paper is to be explained by a certain degree of interaction between the hydrophilic groups of the modified synthetic pulp and the hydroxyl groups of the cellulose. ANIC feels that the addition of synthetic pulp to the furnish of cellulosic pulp has a market in the production of speciality papers, and that packaging materials for liquid and solid foodstuffs (milk, butter, etc.), as well as composite materials consisting of paperboards (providing rigidity) and 100% synthetic pulp sheets (providing gas impermeability) offer interesting potential markets for synthetic pulp.

Montedison

Montedison has developed a synthetic pulp called Ferlosa, based on polyolefins, with an emphasis on Ferlosa C2, based on HD PE (melting point 135°C). But on an experimental basis the company has also developed Ferlosa C3, based on polypropylene, having a high-crystallinity index (melting point 170°C). Contrary to other synthetic pulp processes, the Montedison process involves the incorporation of a finely subdivided mineral

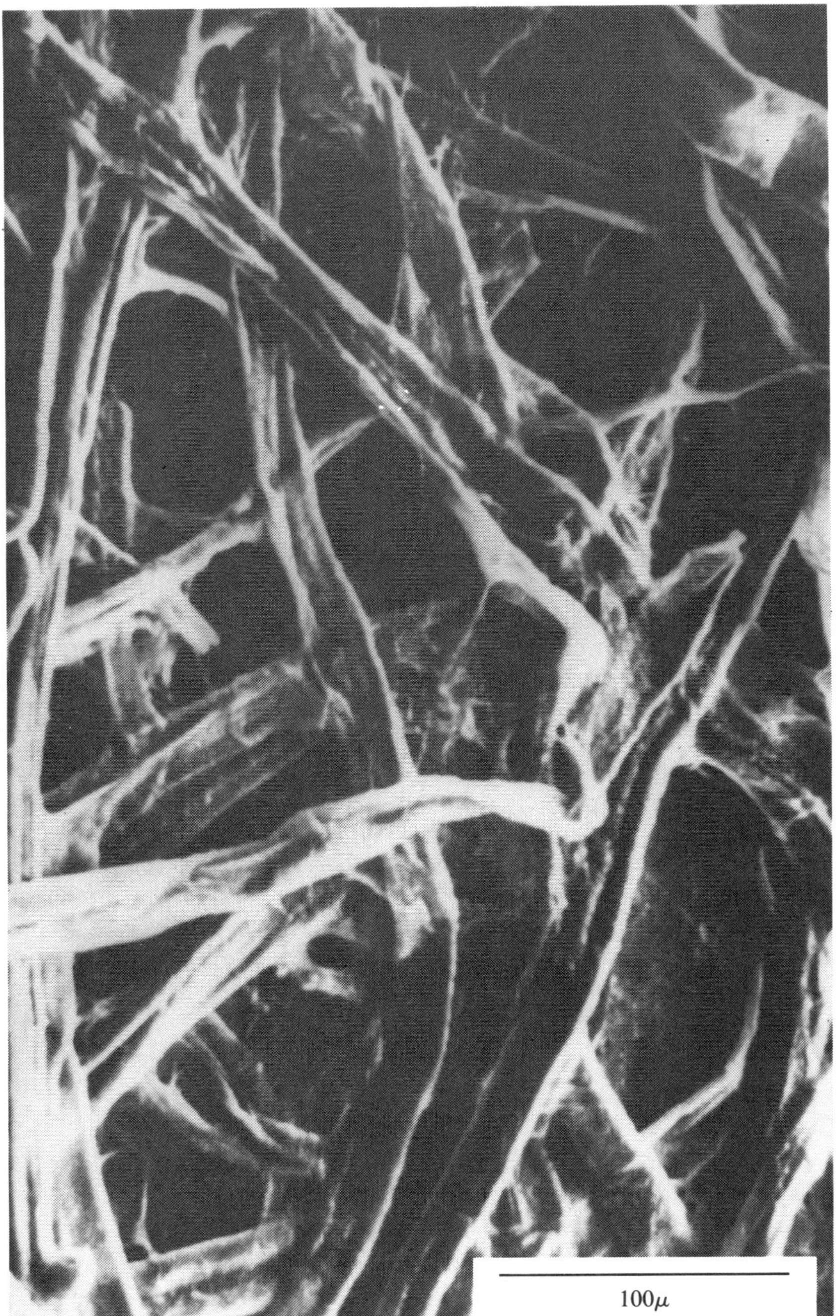
100μ

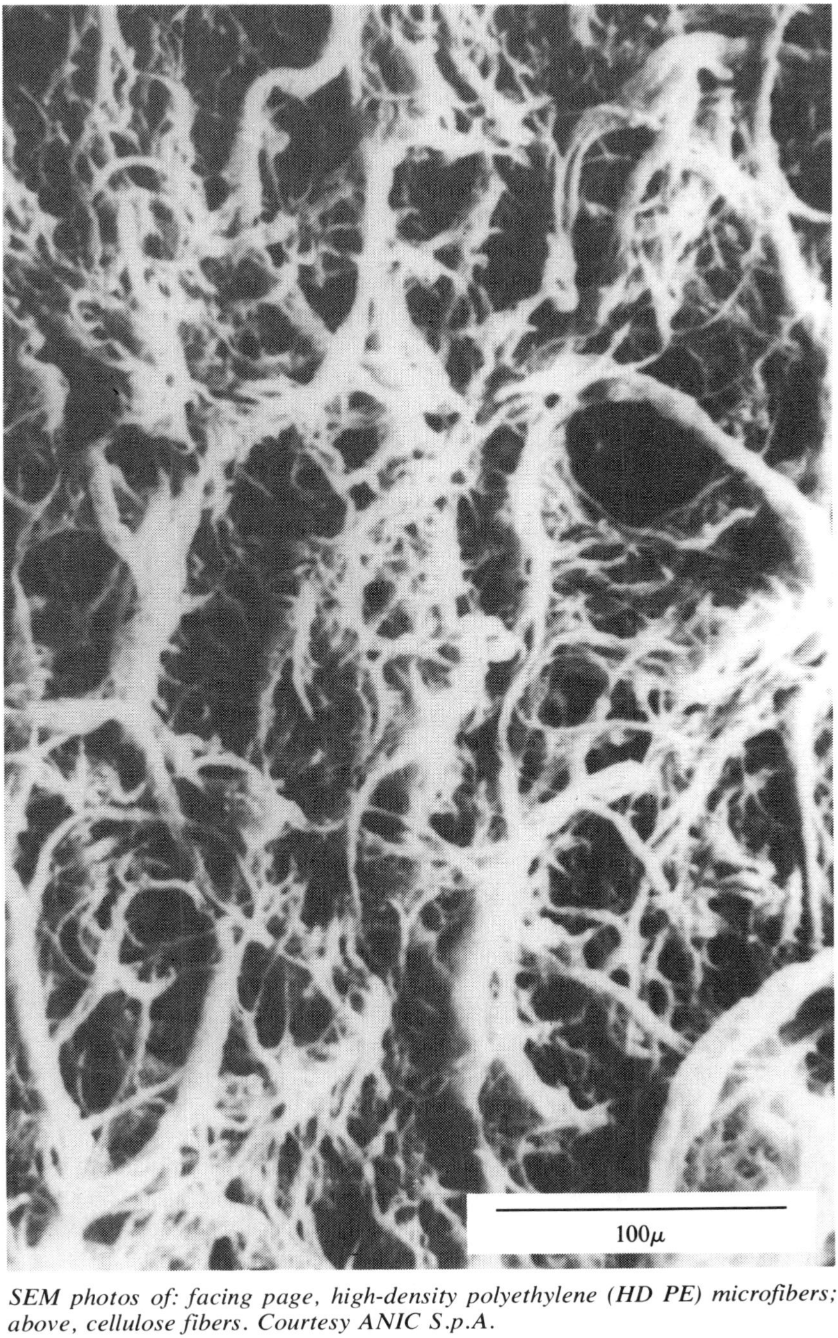

SEM photos of: facing page, high-density polyethylene (HD PE) microfibers; above, cellulose fibers. Courtesy ANIC S.p.A.

Table 4.11 *Properties of Ferlosa Filled and Unfilled Pulp*

Property	Unit	Type of pulp: Ferlosa C2-K-M	Type of pulp: Ferlosa C2-O-M
Fiber length	mm	1.2-1.6	1.4-1.8
Apparent diameter	micron	15-25	15-25
Surface area	m^2/g	5	5
Degree of whiteness	%	86-88	95-97
Ash content	%	24-28	0
Density	g/cm^3	~1.18	0.96
Melting temperature	°C	135	135

filler, which apparently imparts on the pulp "excellent processability properties" and has a positive influence on certain properties of the paper made from it.[13] The main pulp grade produced by Montedison is filled with Kaolin; however, other pigments or mineral or organic fillers may be incorporated in accordance with the application of the pulp. Ferlosa C2-K-M is a HD PE synthetic pulp filled with 30% by weight of Kaolin. The letter *M* in this grade indicates that the pulp fiber is of medium length. For some applications in which the chemical purity of the material is of great importance, Montedison produces Ferlosa C2-O, a type of unfilled pulp. Table 4.11 compares the properties of Ferlosa filled and unfilled pulp.

According to Montedison, properties of polyolefins can be utilized by heat treatment in the production of speciality papers, resulting in the following properties:

- ☐ permanent embossing
- ☐ water impermeability
- ☐ controlled porosity
- ☐ high resistance to moisture
- ☐ sealability
- ☐ transparency
- ☐ high surface cohesion up to film formation

Some of these characteristics are of fundamental importance in the production of quality wallpapers. Montedison has issued a special bulletin on "Ferlosa synthetic pulp in the production of wallpaper" describing first the production of the "support" (with a basic weight of 130-170

13. Montedison, Technical Information, Plastics Division, Bulletin No. 2461 E, February 1976.

g/m^2 on machines with up to 3.5-m widths and operating at a speed of 130 m/min) and then the printing, in which gravure machines of up to six colors are used at a rate of 100 m/min.[14] After the printing, the paper is hot-embossed on machines used for making vinyl wallpapers. Montedison feels that these new wallpapers are highly competitive.

Gulf Research and Development Company

Among firms that have been active in developing synthetic pulp but have not yet announced the commercialization of their developments in this field is Gulf Research & Development Company, of Pittsburgh, Pennsylvania (USA). The firm has been granted several patents related to the production of synthetic pulp and to the production of paper from different blends of natural cellulosic fibers and fibers prepared from linear high-molecular-weight ethylene polymer.[15] Gulf''s invention, called "Improved Paper Product and Process for Production Thereof," relates to paper and paper-like products, particularly to heat-sealable paper.[16]

Among Gulf Research patents related to the production of synthetic pulp are "Improvements in or Relating to a Fibril and a Process for its Manufacture" and "Improved Fibril Process."[17] Both patents refer to oriented polymer solutions generated by centrifugally imparting shear to a hot viscose solution.

The polyolefin fibrils are produced from a high-molecular-weight polymer, and are intended primarily for utilization in the production of paper and other sheet-like materials on papermaking machinery. The firm's invention called "Fibril Process" refers to the method of producing fibrils from a high-molecular-weight polymer (e.g., linear high-density polyethylene) through generating oriented polymer solutions by imparting shear to hot viscose solution by means of die extrusion, wherein the solution is passed through an elongate die under conditions of laminar flow.[18]

Among the latest patents issued to Gulf is "Method of Preparing Fibrils having Properties for making Improved Paper Sheets," which refers to the improvement consisting in incorporating a surfactant into the hot polyolefin/hydrocarbon solvent solution prior to the fibril-

14. Montedison, "Ferlosa Synthetic Paper in the Production of Wallpaper," Technical Information, Plastics Division, Bulletin No. 2460 E, February 1976.

15. See Wolpert, *Plastics and the Paper Industry*, pp. 48-56.

16. British Patent 1,362,344, published August 7, 1974.

17. British Patents 1,372,116 and 1,373,708, published October 30, 1974 and November 13, 1974.

18. British Patent 1,374,640, published November 20, 1974.

formation steps of the process.[19] The resultant fibrils possess improved hydrophilic properties, and paper sheets fabricated from these fibrils possess improved physical properties, particularly increased tensile strength.

STYRENE-POLYMER-BASED SYNTHETIC PULP (TORAY-KANZAKI PROCESS)

Toray Industries, a leading Japanese synthetic fiber manufacturer, and Kanzaki Paper Manufacturing Co., a leading manufacturer of coated and speciality papers in Japan, have been cooperating in the development of new-type-micro-fibrillar synthetic pulp, and in the development of paper from a blend of this synthetic pulp and woodpulp. In October 1974, these two manufacturers founded the joint venture Topka Corporation, Tokyo, for the promotion of the synthetic pulp and the paper project. Toray has a semi-commercial plant with a 1,000 MT/yr synthetic pulp production capacity, which could be easily expanded to a 2,500 MT/yr capacity. Topka has been promoting the sales activities of the synthetic pulp with the intention of establishing a commercial plant after the completion of market developments.

Unlike other manufacturers, whose synthetic pulps are based on polyolefins, Topka has been developing and offering synthetic pulps based on olefin plus styrene polymers and copolymers (also graft copolymers of PP and an additional monomer, e.g., vinyl alcohol), PVA plus acrylonitrile copolymers, and PP vinyl acetate graft copolymers plus PP.

The synthetic pulp developed by the Toray Kanzaki process is claimed by Topka to be so similar to wood pulp in fibrillar structure, as well as in self-bonding property, that it can be formed to paper on conventional paper machines without any special binders or heat-bonding processes. Synthetic pulp can be added to wood pulp at any blending ratio. Topka emphasizes that synthetic paper made from this pulp does not require special disposal, claiming that synthetic pulp waste paper could be reused just as conventional pulp/paper is.

Latest U.S. Patents

The U.S. patent "Synthetic Fiber for Paper and Method for Producing the Same"[20] and the U.S. patent "Method for Producing Synthetic

19. U.S. Patent 3,904,728, Specification published September 9, 1975.

20. U.S. Patent 3,928,469, published December 23, 1975.

Fiber for Paper"[21] both refer to a synthetic fiber that is substantially comprised of:

1. 5-40% by weight of a copolymer consisting of—
   a) 20-80% by weight of polyvinyl alcohol,
   b) 80-20% by weight of acrylonitrile.
2. 60-95% by weight of a copolymer consisting of—
   a) 55-95% by weight of styrene
   b) 5-45% by weight of acrylonitrile.

This indicates the high content of styrene in Toray's synthetic pulp. This fiber is used for producing paper and has a potentially fibrillar texture. When beaten, it forms synthetic pulp having stalk fibrils and micro-fibrils extending from the stalk fibrils. High-tenacity synthetic paper can be made of such fibers because this synthetic pulp is self-adhesive.

The invention is to provide a method of making a fiber for a synthetic paper imparted with performance capable of producing pulp and making paper by exactly the same method normally used for wood pulp. This is done by imparting hydrophilic properties to pulp (a point which has been little considered in the case of conventional synthetic pulp), improving the dispersion properties in water when made into an aqueous slurry, and imparting adhesion strength between the fibrils.

Potential Applications

Topka envisages a wide range of "synthetic pulp blended paper" and has developed the following types of paper—containing synthetic pulp blended with woodpulp—for the following applications:

1. Opaque paper.
   a) Thin paper with high brightness, opacity, and smooth surface for typewriter paper, letterhead paper, airmail paper, super-thin multicolored printing paper, and overlay paper.
   b) Paper with partial transparentization for bond paper.
   c) Paper with irregular design effects, made by embossing, for high-grade wallpaper.
   d) Paper with high stiffness, smooth surface, and high-dimensional stability used for computer paper and for high-grade recording paper.

21. U.S. Patent 3,963,821, published June 15, 1976.

2. Transparentized paper.
   a) Translucent paper for industrial applications (release paper and transfer paper), moistureproof paper, waterproof and oilproof paper.
   b) Transparent paper for high-grade wrapping paper for high-class printing.
3. Translucent paper with rough surface.
   a) To be used on any type of copier for applications for "copy master paper" with clear and sharp reproductions and high copy efficiency.
   b) For drafting paper and high-grade recording paper.

Properties

Table 4.12 lists the properties of machine-made paper containing styrene-polymer synthetic pulp.

Manufacturing Process

The patent called "Method for Producing a Synthetic Paper Forming Material" granted to Toray Industries Inc., Tokyo, relates to a method for producing an oriented paper-forming material containing an olefin or styrene polymer.[22] This method for producing paper-forming material comprises preparing a dispersion from a dispersed phase (which is made up of drops of olefin or styrene homopolymer or copolymer and a solvent compatible with the olefin or styrene polymer, and a dispersion medium) and ejecting the dispersion under pressure through an orifice to obtain solid elongate drops of oriented olefin or styrene homopolymer or copolymer. According to the Patent Specification, paper having good intertwining characteristics, no surface patterns, and excellent appearance, luster, and hand can be produced by using this method, and the paper is uniform and easily printed.

The required orientation of the paper-forming material is initially produced by ejection, which elongates the dispersed phase by sheer forces and thus orients the polymer in the dispersed phase. The size and length of the paper-forming fibers produced are determined by the particle size of the drops from 3 to 400 microns. Most of the paper-forming material obtained by this method preferably consists of fibers having a thickness of 10 to 1,000 microns, and a length of 50 to 25,000 microns. The inside of the respective fiber is a structure in which fibrils, ribbons or films whose diameter or thickness is not more than 10 microns, are

22. British Patent Specification 1,323,174, published July 11, 1973.

Table 4.12 *Properties of Machine-Made Papers Containing Styrene-Polymer Synthetic Pulp*

Property	*Opaque-type paper*	*Transparent-type paper*
Synthetic pulp content (%)	20	25
Basic weight (g/m^2)	39.7	40
Thickness (micron)*	50.0	36.0
Density (g/cm^3)	0.79	1.12
Expansion in water (%)	1.3	0.8
Opacity (%)	81.4	31.0
Brightness (%)	82.5	65.7
Breaking length (km)		
MD	5.0	8.1
TD	4.6	5.2
Folding endurance		
MD	45	900
TD	70	1,370

* Thickness calculated by basic weight and density.

Source: Toray/Topka.

assembled or dispersed integrally in three dimensions, and may be compressed to produce a flat cross section. The paper strength is developed by intertwining the paper-forming material as a whole in a three-dimensional manner and by contact within the paper forming material.

The Patent Specification says that polyethylene, polypropylene, or an ethylene-propylene copolymer is especially preferred and that good results are also obtained using a graft copolymer of PP and an additional monomer, such as vinyl alcohol, vinyl acetate, or acrylic acid in an amount not exceeding 30% by weight of the polypropylene. The additional monomer may be water soluble. The process for producing paper from the paper-forming material may use a coarse screen.

A very simple method is to make paper from fibrous sheets collected on wire transfer netting (e.g., on a conveyor belt). The paper-forming material can be ejected not only on the wire netting but also directly on or in a fluid such as water; thereafter, the jetted paper material may be hit and smashed or another component may be added, and then a paper may be made from the material. The paper material may be also blown onto a film to make a laminated paper. If required, a natural pulp or a synthetic fiber may be blended in. Paper made from this material may also be produced by a dry method.

The Patent Specification includes 32 examples of this patent's application. These examples included data on synthetic paper made from

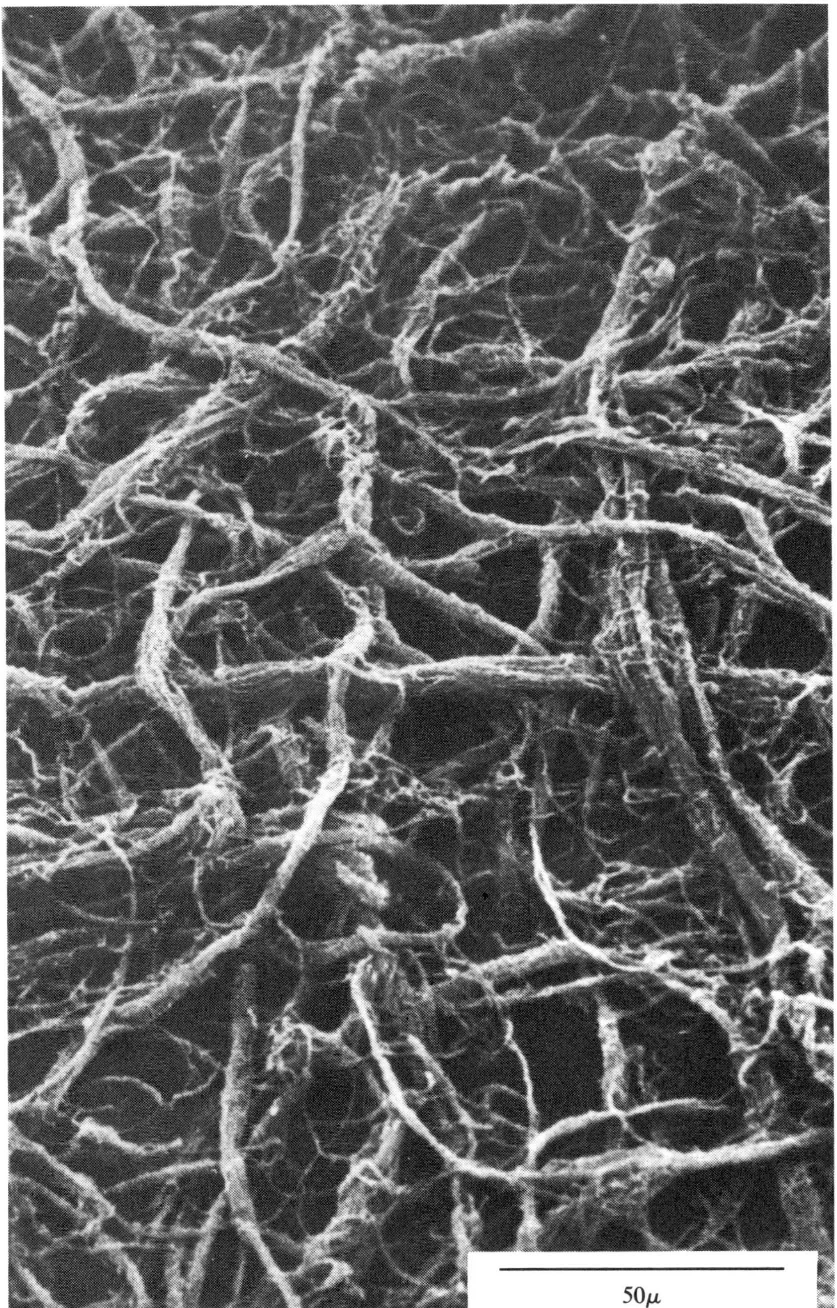
50μ

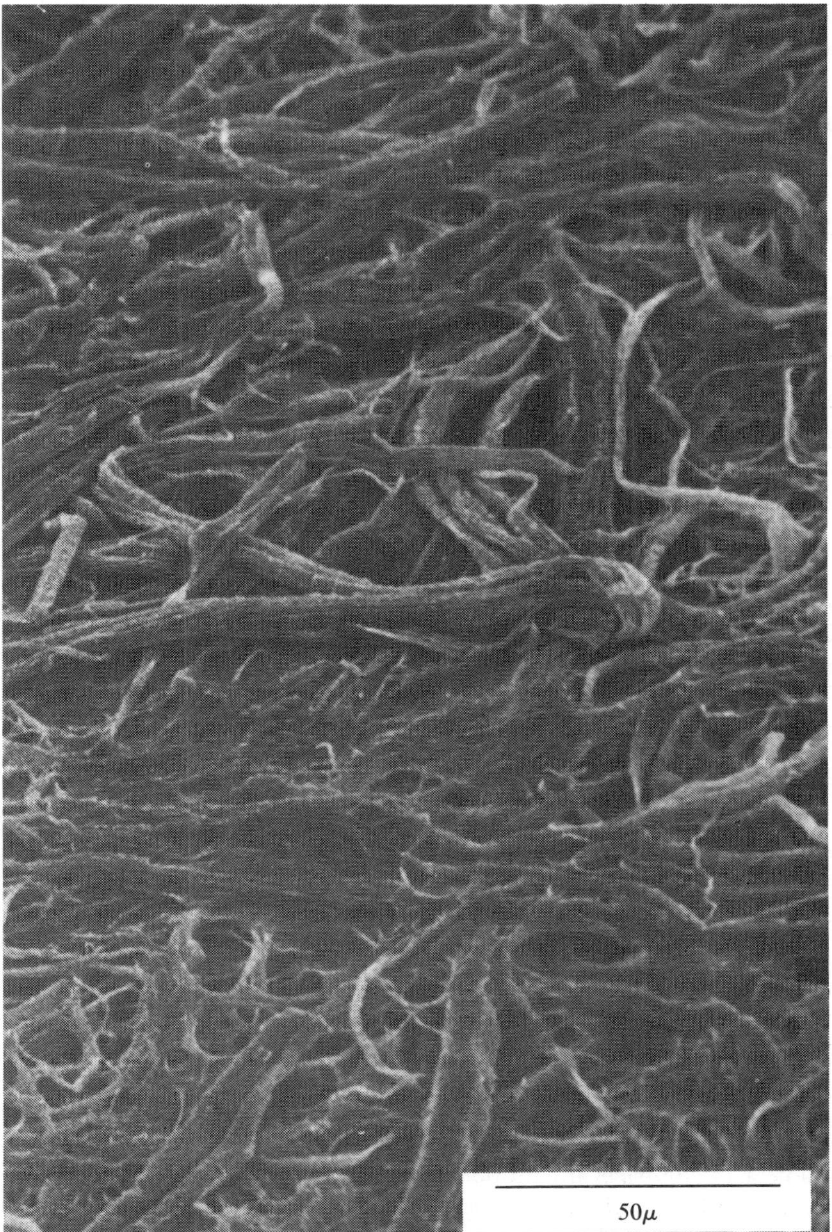

SEM photos of: facing page, styrenic-polymer-based synthetic pulp; above, paper containing 20% weight of styrenic-polymer-based synthetic pulp. Courtesy Toray Industries.

Table 4.13 *Data on Synthetic Papers Made from Various Polymers*

Polymer	*Paper-forming material* thickness (micron)	*Paper-forming material* length (micron)	*Basic weight of paper (g/m²)*	*Tenacity of paper (kg/cm²)*
Polypropylene	1,000	10,000	180	3.2
HD PE (Mitsui Petrochemical)	1,000	10,000	126	0.6
10% PE block co-polypropylene	1,000	20,000	98	1.3
7% Vinyl acetate block co-polypropylene	1,000	20,000	88	1.3
6% acrylic acid graft PP	1,000	20,000	93	1.2
Polybutene-1	1,000	5,000	110	0.3
Polypropylene	1,000	10,000	97	2.8
Polypropylene	1,000	15,000	125	1.5
Polypropylene	1,000	20,000	111	1.3
Polypropylene	600	10,000	113	1.2

Source: Synthetic and Plastics Paper, Vol. 4, p. 713. Compiled from British Patent 1,323,174 Complete Specification published July 11, 1973.

different polymers. In all examples water is listed as the dispersion medium, and the comments say "paper-like matter of uniform texture." Table 4.13 contains the data on synthetic papers made from different polymers.

The patent called "Process for Making Paper Forming Pulp" describes compositions of hydrophobic polymers with hydrophilic segments.[23] This new patented process is based on spinning the fiber, and the filaments are cut to required length. This synthetic paper has, according to the Patent Specification, high mechanical strength, especially tensile and flexural strength, and may be used for bank notes, posters, maps, electrical insulation paper, wallpaper, filter paper, nonwoven fabric, reinforcing agents for plastics, and much else.

Polyolefin-based Synthetic Pulp

Toray has also started developing synthetic pulp on a polyolefin basis, but this development has not yet reached the pilot plant stage. Table 4.14 compares the properties of paper containing 30% polyolefin-based synthetic pulp with those of paper (after size press) made from a furnish of softwood and hardwood bleached kraft pulp.

23. British Patent 1,357,817, Complete Specification published June 26, 1974. See Wolpert, *Plastics and the Paper Industry*, pp. 36-39.

Table 4.14 *Comparison of Properties of Paper Containing Polyolefin-based Synthetic Pulp with Wood Pulp Paper*

Property	*Paper containing 30% synthetic pulp*	*Wood pulp paper*
Basic weight (g/m^2)	51.2	50.8
Thickness (micron)*	75.0	73.0
Density (g/cm^3)	0.68	0.70
Brightness (%)	83.6	79.6
Opacity (%)	84.8	70.1

* Thickness calculated by basic weight and density.

Source: Toray/Topka.

Improved Surface Strengths

Remarkably improved surface strengths of paper sheets composed of synthetic fibrous materials together with cellulose pulps are claimed in the recent U.K. patent "Fibrous Synthetic Paper-Forming Materials", granted to Toray Industries (Japan).[24] Polymers referred to in this invention have a hydrophilic component (e.g., an alkali metal salt of a carboxylic acid group) and a lipophilic component (e.g., a polymer chain or hydrocarbon chain) and therefore function as surface active agents. The patent relates to "methods of making paper-forming fibrous materials, to the products of such methods, and to paper made therefrom."

The method of making a paper-forming fibrous material is by preparing an emulsion in which the disperse phase is a solution of one or more poly-alpha-olefins in an organic solvent, and the dispersion medium is a nonsolvent for the poly-alpha-olefins, is substantially incompatible with the organic solvent, and contains a surface acting agent. "The surface active agents used in the method of the invention not only act as emulsifiers but also contribute hydrophilic and self-bonding properties to the resulting paper-forming materials."

Polyethylene, polypropylene, poly-1-butene, poly-4-methyl-1-pentene and copolymers of two or more alpha-olefin monomers are described as suitable poly-alpha-olefins for use in the method of this invention.

As the dispersion medium the use of water is preferable.

"Preferably the solution of the poly-alpha-olefin is dispersed in small particles in the dispersion medium, and the dispersion medium and the solvent are preferably present in the ratio of from 1:6 to 6:1 by volume."

24. British Patent 1,440,737, Complete Specification published June 23, 1976.

In the method of the invention the preparation of the emulsion is preferably at a temperature which is sufficient to dissolve the polymers in the solvent and then to remove the solvent by vaporization from the polymer during the ejecting.

The ejection of the emulsion is sometimes referred to as a jet-extrusion or jetting process or an emulsion flush-spinning process, and can be conveniently carried out at autogeneous pressure or at a pressure above autogeneous pressure to obtain a high production rate.

The particle size of the emulsion varies; it may be from 0.1 to 100 microns. The required orientation of the paper-forming materials is initially produced by expansion extrusion which elongates the dispersed phase by shear forces and the polymer in the dispersed phase is molecularly oriented and intertwined. Subsequent to jet-extrusion, the polymer must be solidified quickly before undesirable relaxation of it occurs.

After the synthetic paper-forming fibrous material is formed into paper, using if desired an internal sizing reagent, the resulting paper sheet may be subjected to size press coating with the surface size press as follows: "After the fibrous material of the invention is disintegrated using an apparatus such as a domestic mixer, a PFI mill, a refiner or a beater, *the resulting fibrous pulp-like material is mixed with natural pulp and formed into paper containing both pulps*" (italics mine).

In the case of internal sizing of the synthetic fibrous material or a mixture with natural pulp, the following processes can be applied:

1. Producing a pulp which is treated with an internal sizing agent after mixing a synthetic fibrous material and a natural pulp.
2. Producing a pulp which is obtained by mixing a natural pulp with a synthetic fibrous material previously treated with an internal sizing agent, and, if desired, treating the resulting pulp mixture with an internal sizing agent.
3. Producing a pulp which is obtained by mixing a natural pulp with a synthetic fibrous material that has been disintegrated in a solution of an internal sizing agent, and, if desired, treating the resulting pulp mixture with an internal sizing agent.

The resulting wet sheet is subjected to dehydration and drying, and afterwards it is preferably coated with at least one surface size press coating agent, in order to produce a paper sheet which has a higher surface-picking strength. The amount of the size press agent is suitably 0.1 to 5 g/m^2, based on the weight of the paper sheet.

The list of examples of preferable compositions of polymer mixture and surface active agents shows that in most cases polypropylene and various styrene copolymers are recommended.

Synthetic Fibrous Material in Support Sheet for Pressure Sensitive Recording Systems

In pressure sensitive recording systems, lightweight natural fiber papers of 30 to 50 g/m^2 have been used. The recently patented joint invention of Kanzaki Paper and Toray Industries (Japan) claims that their invented support sheets have superior functional properties (rate of tetralin penetration, brightness, opacity, dimensional stability to moisture, smudge of pressure characteristics) compared with the lightweight natural fiber papers.[25]

The invention relates to two basic types of new support sheets having a coefficient of tetralin penetration rate of 600 x 10^{-3} or less:

1. Comprising fibrous materials, at least a part of which is a synthetic fibrous material.
2. Made of a "modified cellulose fibrous material," namely—
   a) A fibrous material obtained by carrying out the polymerization of at least one vinyl monomer on or within natural cellulose fibers. The polymerization of the vinyl monomer is effected in the presence of the cellulose fibers, and when these fibers are hollow or porous, at least some of the polymerization occurs within the cellulose fibers.
   b) A fibrous material obtained by depositing a homopolymer or copolymer of at least one alpha-olefin on natural cellulose fibers.

The patent specification says that "Preferably the support sheet of the paper material according to the invention is made of a fibrous material comprising a mixture of natural cellulose fibres and synthetic fibres, although the whole support sheet may be made of synthetic fibres alone."

Useful synthetic fibers can be selected from widely varying groups of polymers. The synthetic fibrous material preferably contains therein at least 50% by weight of the units of one or more alpha-olefins. The three preferred embodiments of the invention are:

1. Polyolefin synthetic fibers (having a hydrophilic component) containing at least 50% by weight of units of one or more alpha-olefins are used for forming the support sheet. The spinning material consists essentially of—
   a) An alpha-olefin homopolyer or a copolymer.

25. British Patent 1,435,016, Complete Specification published May 12, 1976.

b) A hydrophilic or water-soluble polymer component (e.g., polyvinyl alcohol). The polymers may be spun by various spinning techniques to produce fibrous material, and the material obtained by the emulsion flush spinning may be subjected to beating and/or refining if required. The fibrous material obtained in this way is easily fibrillated, has a relatively high specific surface area, is a good oil absorbent, and is dispersive to water.

2. The synthetic fibrous material contains at least 60% by weight of vinyl alcohol units. The polyvinyl alcohol resin spinning material may be obtained by hydrolizing a homopolymer of vinyl acetate or a copolymer of vinyl acetate with a monomer such as vinyl chloride, an acrylic ester, or a methacrylic ester. The spinning material comprising polyvinyl alcohol (and which may contain other water-soluble polymeric material, e.g., starch, casein) may be transformed by any conventional spinning technique to produce fibrous material which may be subjected to a beating treatment for fibrillation.
3. The fibrous material comprises the modified natural cellulose fibrous material *or* the fibrous material obtained by depositing a homopolymer or copolymer (comprising at least one alpha-olefin monomer) on synthetic fibers. In the case of a modified, natural cellulose fibrous material the vinyl monomer polymerized on or within the natural cellulose fibers is preferably one or more of the following: vinyl acetate, acrylic acid, acrylic ester, acrylamide, methacrylic acid, methacrylic ester, and methacrylamide. The modified natural cellulose fibrous material may be fibrillated by a beating treatment if required.

The patent specification says that "Of the three preferred embodiments described, the first embodiment is particularly preferred. Poly-alpha-olefin fibres obtained by the emulsion flush spinning techniques would be the best." The support sheet having a coefficient of oil penetration rate below the upper limit is formed of various synthetic fibers alone or of a mixture of synthetic fibers with other fibrous materials, including natural cellulose fibers (most economically, bleached woodpulp is used), and the formation of the sheet may be carried out by either wet or dry sheet-forming techniques.

FIBRILLATION TECHNOLOGY

This technology has developed from the introduction in the mid-sixties of polyolefin stretched plastics tapes made from extruded polypropylene or high-density polyethylene film. These tapes have found a

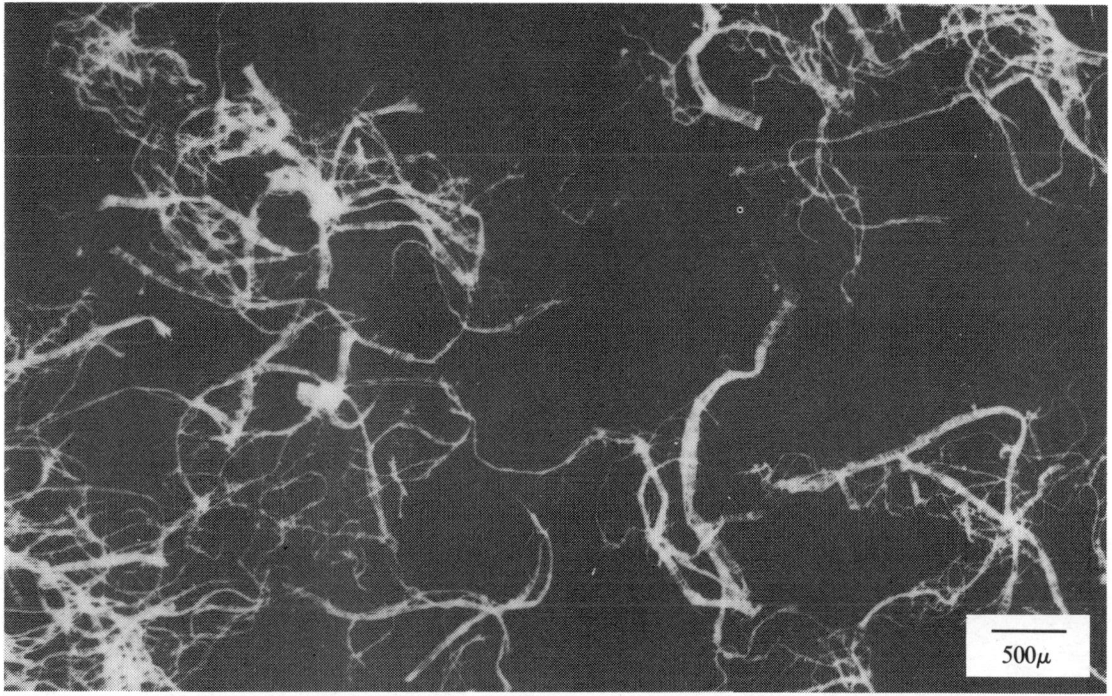

Danpulp is made from Ube Nitto Kasei of Japan's Danchip, which is mainly a polyolefin resin with a branching structure similar to woodpulp. Compare this SEM photo with the one of bleached kraft pulp after beating on the following page. Courtesy Ube Nitto Kasei.

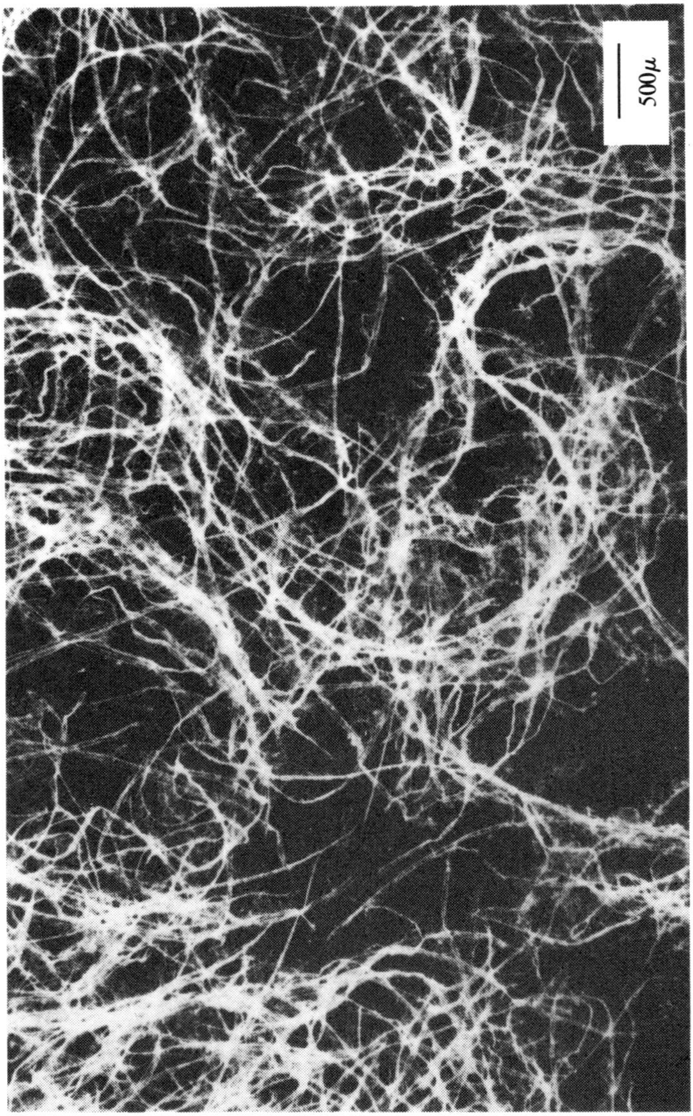

This SEM photo shows bleached kraft pulp after beating. Compare this photo with the one of Danpulp on the preceding page. Courtesy Ube Nitto Kasei.

good market in carpet backing, wool packs, sacks, rope and twines, etc., mainly as a replacement for jute and sisal.[26]

The next step of development was to fibrillate (break-up the tape or film into fibrils) the oriented tape or film by further stretching (orientation) and then use the fibrillated material either as staple fiber or continuous yarn for various textile applications, including polypropylene carpet-face yarn.

Although in some textile applications, fibrillation technology has proved to be an economically viable proposition (as against the traditional spinning technology) it is felt that for the paper industry this process is more expensive than the "from monomer to pulp" (polymerization of resin) process. However, it is possible that for some applications this type of synthetic pulp will find a market.

Ube Nitto

Ube Nitto Kasei Co. Ltd., Tokyo, has developed DanPulp, which is produced by fibrillization of DanChip, made mainly from polyolefin resin. DanPulp is claimed to have a branching structure similar to wood pulp and to be suitable for sheet formation by the wet process. Ube Nitto feels that this synthetic pulp can be used immediately as fillers for asbestos products and wallpaper. Work is also in progress to use DanPulp for synthetic leather bags, moulded paper vessels, etc. Because of its inertness, DanPulp can be used for making oil filters and for similar applications. Tables 4.15 to 4.19 list the properties of paper made from DanPulp/LBKP.

Sumitomo Chemical

A different method of fibrillation technology is the process by Sumitomo Chemical Co. Ltd., Osaka. The company's short fibrous material—TPS fiber—is claimed to have unique branches and wide distribution of fiber length. The fiber can be made in 1- to 15-mm lengths and 0.005- to 0.1-mm widths. The process involves uniaxial orientation of isotactic PP, LD PE, HD PE film, which is cut into pieces of 2-15 mm; these pieces are beaten or refined in water—in this way fibrillating hydrophobic polyolefin in water into fine fibers for papermaking.

Paper made from 100% TPS fiber requires a suitable binder. TPS fiber can be mixed with cellulose pulp and then made into sheet, which by heat treatment results in a product of high wet strength. This process can

26. See Vladimir M. Wolpert, *Plastics Tapes for Textiles*, 2 vols. (Tenterden, Kent, England: Marketing Man-Made Materials, 1968).

Table 4.15 *Properties of the Paper Made of Danpulp/LBKP*

Property				*No binder*			
Danpulp content (%)	0	15	30	0	30	50	80
Basis weight (g/m^2)	50	50	52	75	77	73	73
Moisture content (%)	6.6	5.2	4.5	10.3	5.5	4.2	2.1
Bulk density (g/cc)	0.84	0.80	0.77	0.65	0.42	0.39	0.35
Brightness by Hunter (%)	79.8	82.2	83.5	83.2	85.1	86.8	89.4
Opacity (%)	79.6	81.1	82.6	79.5	91.6	91.4	92.5
JIS Mullen burst factor	1.5	0.9	1.0	1.5	1.0	0.8	0.7
Breaking length (km)	3.6	3.6	2.4	3.5	1.9	1.5	0.9
Elmendorf tear factor	64	88	115	64	78	66	63
MIT folding endurance (cycle)	35	12	6	30	2	2	2
Wet breaking length (km)	0.25	0.24	0.23	–	–	–	–
Elongation in water (%)	2.15	1.10	0.30	2.2	–	0.5	0.3
Dimensional change by relative humidity							
Shrinkage (%)	0.258	0.195	0.200	–	–	–	–
Elongation (%)	0.161	0.146	0.152	–	–	–	–
Gurley permeability (sec)	20	15	15	12	–	3	3
Smoothness (mm Hg)	41.1	39.5	39.1	–	–	–	–
Stiffness (mg)	–	–	–	–	238[a]	184	103
Young's Modulus (dyne/cm^2) x 10^{10}	3.27	2.80	2.02	–	–	–	–
Friction factor	0.645	0.604	0.471	–	–	–	–
Heat-curl factor (degree)	29.8	26.2	6.5	–	–	–	–
Wet-curl factor (degree)	35	10	flat	–	–	–	–

Source: Ube Nitto.

a. Gurley method
b. Clark method
c. NBKP

Table 4.15 *Continued*

Property	*Original*	*With binder*	*Original*	*Heat calendering (180° C)*	*Original*	*Heat calendering (130° C)*
Danpulp content (%)	80	80	50[c]	50[c]	100	100
Basis weight (g/m^2)	100	125	121	131	108	114
Moisture content (%)	1.4	0.8	–	–	0.3	0.3
Bulk density (g/cc)	0.58	0.64	0.46	0.55	0.39	0.64
Brightness by Hunter (%)	84.4	79.6	82.8	75.9	91.7	83.8
Opacity (%)	97.8	95.8	93.8	80.6	88.5	75.9
JIS Mullen burst factor	0.8	1.9	2.1	3.4	0.6	1.0
Breaking length (km)	1.0	2.3	2.3	3.2	0.4	2.4
Elmendorf tear factor	78	110	160	128	112	196
MIT folding endurance (cycle)	2	1,160	158	1,430	–	22
Wet breaking length (km)	0.2	0.4	0.3	0.8	0.2	2.1
Elongation in water (%)	0.8	0.9	0.5	0	0	0
Dimensional change by relative humidity						
Shrinkage (%)	–	–	–	–	–	–
Elongation (%)	–	–	–	–	–	–
Gurley permeability (sec)	90	190	4	48	5	136
Smoothness (mm Hg)	–	–	(sec) 1.5	2	–	800
Stiffness (mg)	24[b]	23	356[a]	425	74	138
Young's Modulus (dyne/cm^2) x 10^{10}	–	–	–	–	–	-
Friction factor	–	–	–	–	–	–
Heat-curl factor (degree)	–	–	–	–	–	–
Wet-curl factor (degree)	–	–	–	–	–	–

Table 4.16 *Relation between DanPulp Fiber Lengths and Properties of the Paper (Sheets Made of 100% DanPulp)*

DanPulp: Average fiber length (mm)	DanPulp: Average fiber diameter (μ)	Treatment of heat-calendering machine	Basis weight (g/m^2)	Thickness (mm)	Bulk density (g/cc)	JIS Mullen burst factor	Breaking length (km)	Elmendorf tear factor	Wet breaking length (km)	Elongation in water (%)
1.2	30	none	138	0.43	0.32	0.4	0.15	–	0.1	0.5
1.2	30	140°C (both sides)	143	0.21	0.68	0.9	2.2	105	2.2	0
7.8	45	none	179	0.50	0.36	0.7	1.1	320	0.3	0
7.8	45	140°C (both sides)	182	0.24	0.76	1.6	2.3	320	1.4	0
15.7	45	none	170	0.48	0.35	1.0	1.3	690	0.4	0
15.7	45	140°C (both sides)	165	0.27	0.61	1.9	2.3	550	2.0	0

Source: Ube Nitto.

Table 4.17 *Uses of the Paper Made of DanPulp/LBKP as Filtering Material*

Filters	Thickness of sheet (μ)	Basis weight (g/m^2)	Bulk density (g/cc)	Mullen burst (kg/cm^2) strength: dry	Mullen burst (kg/cm^2) strength: wet	Air (mm Hg) permeability	Filtering resistance $\left(\frac{mm\ Hg/cm^2}{cm^3/sec}\right)$
Paper made of DanPulp/LBKP (80/20)	280	100	0.36	0.6	0.25	66	0.20
Toyo filter paper No. 5c	200	100	0.5	1.2	0.1	55	0.43

Source: Ube Nitto (quoted in Wolpert, *Synthetic and Plastics Paper*, Vol. 4, 1974, p. 718).

Table 4.18 *Oil Absorbent Quality of DanPulp Paper*

Sample			*Oil absorbed*			
DanPulp content (%)	*Basis weight (g/m^2)*	*Thickness (mm)*	*Turbine oil (%)*	*Turbine oil on water (%)*	*C-machine oil (%)*	*C-machine oil on water (%)*
100	170	0.48	161	162	196	169
40	145	0.42	168	157	191	168
0	150	0.30	108	64	132	94

Source: Ube Nitto.

Table 4.19 *Weathering Test of Paper Made of DanPulp/NBKP (80/20)*

	Exposure time to Sunshine-weatherometer (hr)					
	0	*80*	*160*	*240*	*300*	*400*
Brightness by Hunter (%)	85.5	81.4	81.4	81.0	80.6	80.5
Tensile strength (kg)	4.18	4.14	3.98	3.88	3.60	3.61
Tensile strength maintained (%)	100	99	95	93	86	86

Source: Ube Nitto.

Table 4.20 *Physical Properties of Papers from Polyolefin and Pulp*

Testing item	*Testing method*	*Unit*	*PP 100%*	*HDPE-NBKP 50%-50%*	*NBKP 100%*
Thickness	JIS P8118	mm	0.4	0.1	0.1
Basic weight	JIS P8124	g/m^2	290	45	53
Elongation	JIS P8113	%	11	12	3
Breaking length (dry)	JIS P8113	km	2.6	3.2	3.6
Breaking length (wet)	JIS P8115	km	2.4	1.1	0
Tear factor	JIS P8116		125	115	107
Burst factor	JIS P8112		2.1	1.1	2.7

be used for making a wide range of products, from high-quality writing paper to paper board.

Table 4.20 lists the physical properties of paper manufactured from 100% polypropylene, 50% HD PE and 50% NBKP, and 100% NBKP (polymers in the form of TPS fiber).

Hitachi-Jujo

Another fibrillation process for making synthetic pulp, and a further example of cooperation between a synthetic polymer company and a paper company, is the Hitachi Chemical Company-Jujo Paper Company process. The product is a polyolefin pulp of high fibrillation. The fibers can entangle one another, and the pulp has sufficient wet strength, is readily available for papermaking on conventional equipment, and can be mixed at any desired ratio with natural pulp. Paper made from this pulp has the trade mark Nouvela.[27] According to "Synpulp Projects and Technologies" the second Hitachi-Jujo process is reported but not confirmed to be a reactor process.

Biax-Fiberfilm

Biax-Fiberfilm Corp. (Neenah) has lately been active in development work in the field of fibrillation technique, using blends of incompatible polymers or the incorporation of fillers into polymer compounds, and applying orientation (stretching) methods with higher-production rates

27. In "Synpulp Projects and Technologies" (see footnote 2), Dr. William H. Hoge includes the text of the joint article by Masao Asaoka, Hitachi Chemical, and Koichi Nagai, Jujo Paper, "Synthetic Pulp and Synthetic Pulp Paper," *Japan Plastics Industry Annual* (June 1973). "Synpulp Projects and Technologies" also includes the condensed translation of the West German Patent 2,141,025 published February 24, 1972, "Process for the Preparation of Pulp-like Materials using Plastics as Raw Materials," granted to Hitachi. Masao Asaoka is listed first among several inventors in this Patent.

than those commercially obtained at present.[28] Biax-Fiberfilm aims its monoaxial-stretching operation (the separation of fibrils is then carried out by mechanical means) at the synthetic pulp market, and its biaxial-stretching operation at the synthetic paper market. The application of special types of Kaolin in polymer compounds from which oriented materials are produced has been described in Chapter 3.

PAPER-LIKE POLYOLEFIN SHEET FROM POLYOLEFIN MICRO-FLAKE AGGREGATION

Mitsubishi Rayon Co. Ltd., Tokyo, has patented a process for making a polyolefin sheet from polyolefin micro-flake aggregation.[29] This patent refers to a polyolefin (PP, PE, or mixture of PP and PE) micro-flake aggregation useful for manufacturing paper-like synthetic sheets; the aggregation is easily micro-flakable into a plurality of micro-flakes of various shapes with an average thickness of 5 to 20 microns.

This process for manufacturing the polyolefin micro-flake aggregation involves the preparation of a solution by dissolving 5% to 40% by weight of at least one polyolefin at a temperature not lower than 100°C in a solvent having a boiling point not higher than the temperature of the solution, and then ejecting this solution from a closed container into the atmosphere through an orifice.

This process of forming a paper-like synthetic sheet from this aggregation involves two steps:

1. Micro-flaking away the polyolefin micro-flake aggregation into a plurality of micro-flakes in an organic liquid which is capable of swelling or partially dissolving polyolefin at a temperature not higher than 60°C.
2. Homogeneously suspending micro-flakes in organic liquid, forming suspended micro-flakes into a paper-like sheet, and evaporating the organic liquid from the resultant sheet.

POLYVINYL ALCOHOL FIBER FOR PAPERMAKING

Kuraray Co. Ltd. (formerly Kurashiki Rayon Co. Ltd.), Osaka, manufacturer of Kuralon, a polyvinyl alcohol fiber, has developed special grades of polyvinyl alcohol fiber for papermaking. These grades are

28. Dr. E. Schwartz, BiaxFiberfilm Corp., "Fibrillating Polymer Blends" (Presentation at the TAPPI International Synthetic Papers & Pulps Symposium, Washington D.C., March 31-April 2, 1976).

29. British Patent Specification 1,262,531, published February 2, 1972.

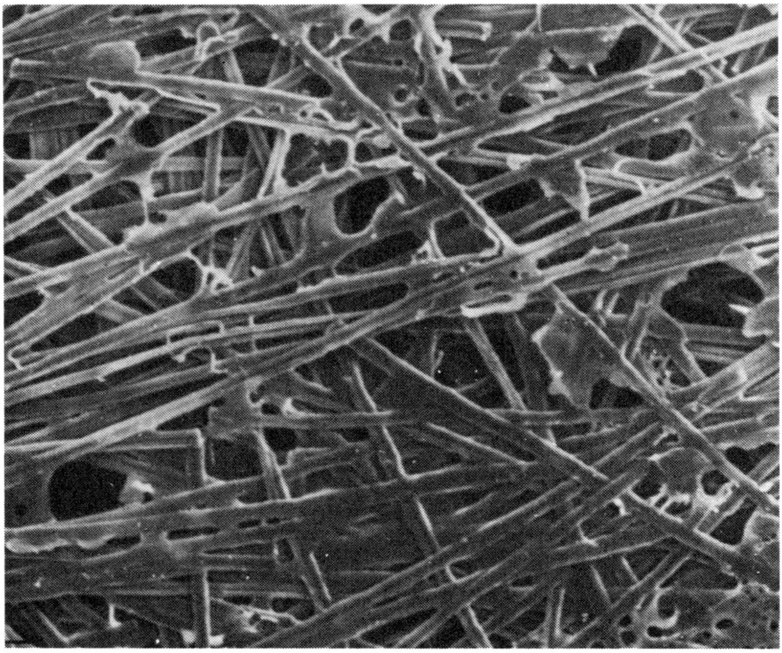

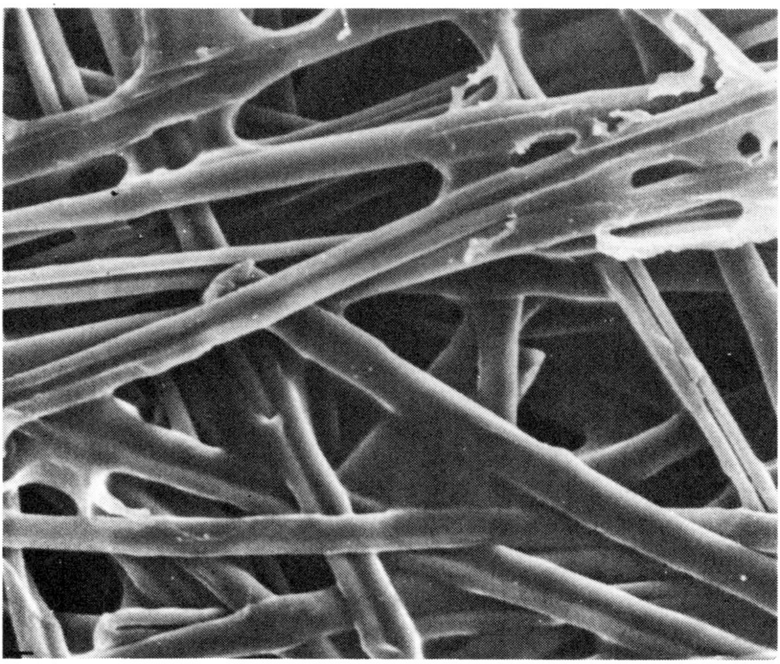

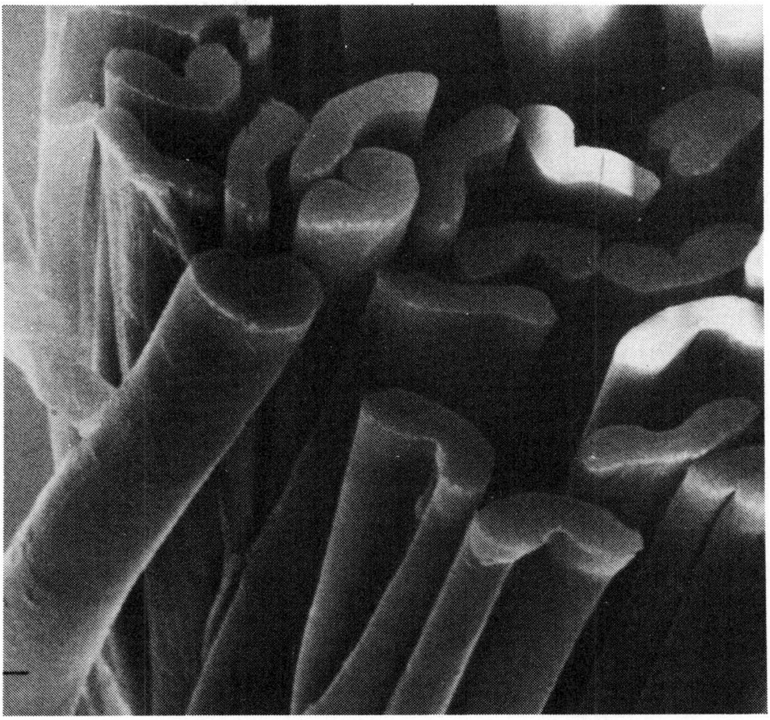

Kuralon VP fiber from Kuraray Co. Ltd. of Japan is a special polyvinyl alcohol fiber which can be prepared as an easily-soluble or hardly-soluble fiber. It can be used as a binder in man-made fiber paper, as a reinforcing material for cellulose-based paper, or as a bonding agent for multi-ply board. SEM photos show this product at: top of facing page, 100x; bottom of facing page, 300x; and above, 1500x. Courtesy Kuraray.

marketed under the trade name Kuralon VP, which is supplied in various easily-soluble and hardly-soluble grades.

Because of its high adhesive property, the easily-soluble VP can be used as a binder in the manufacture of man-made fiber paper, a reinforcing material of pulp paper, or an agent to prevent the separation of ply from the paper board. The easily-soluble VP fiber of long length has also been used as a binder in nonwoven fabrics.

The hardly-soluble VP is hydrophilic and suitable for blending with cellulosic fibers (and also with synthetic fibers) to reinforce the strength of the paper made from these materials. It is also possible to make paper consisting solely of this fiber.

The easily-soluble VP fiber is claimed to display an excellent binding effect in Kuralon and rayon paper, and has been used for this application in Japan. This fiber is spun from polyvinyl alcohol by wet-spinning technology. It shows a high degree of swelling and becomes viscous on the surface, developing a strong adhesive property in water at room temperature. It acts as a binder without losing its fiber form. When this fiber is dispersed in water and blended with other papermaking materials, it absorbs water, swells, and becomes sticky. When subjected to a dryer temperature slightly higher than the dissolving temperature of the fiber itself (in the presence of a specified water content), the adhesive action of the easily-soluble VP fiber results in its bonding with other fibers.

The hardly-soluble VP fiber can be used for making paper because of its high hydrophilic property and strength. When blended with other fibers it improves the tear strength and folding property of the finished paper.

Conventional papermaking machines can be used for processing this fiber. However, the fiber has a tendency to flocculate, which makes it difficult to produce paper of smooth surface. Kuraray has therefore developed a dispersing agent which results in a satisfactory dispersion of the fiber.

According to Kuraray the Kuralon hardly-soluble VP fibers have the following characteristics:

- □ Paper made from these fibers has a high tensile strength, as the fiber itself has a high tenacity, and its adhesiveness to Kuralon VP binder is "excellent."
- □ The cross section of fiber is flat, which helps to increase the strength of paper.
- □ It has excellent weather resistance.

The standard properties of Kuralon VP are listed in Table 4.21.

Kuraray reported that consumption of its VP fiber in Japan increased from 1,250 MT in 1968 to 2,180 MT in 1971 to 2,700 MT in 1973.[30] In addition, their exports of VP fiber reached 360 MT in 1973. In 1974, because of the recession, the consumption of VP fiber for papermaking in Japan dropped to 2,200 MT and the exports of that fiber dropped to 280 MT.

Paper made from a mixture of hardly-soluble and easily-soluble VP fibers is manufactured by Hirose Paper Manufacturing Co. and marketed under the trade name Papylon.

30. See V.M. Wolpert, "Use of Short-Cut Rayon and PVA Fiber for Papermaking Increases in Japan," *Pulp & Paper International*, 16, no. 9 (August 1974), p. 43.

Table 4.21 *Standard Properties of Kuralon VP*

Property	*103 Wet*	*102 Wet*	*101 Dry*	*105-1 Dry*	*105-2 Dry*
Denier (dr)	1.0	0.97	1.5	1.1	1.1
Dry tensile tenacity (g/dr)	7.0	7.2	5.0	3.4	3.4
Dry tensile elongation (%)	13	16	16	25	30
Shrinkage in water					
at 20°C (%)	1	1	20	30	40
at 50°C (%)	1	2	35	40	45
Maximum shrinkage in water (%)	0	10	60	50	50
Dissolving temp. in water (°C)	>100	99	80	70	60
Commercial moisture regain on the weight of bone-dry PVA fiber (%)	5	5	5	5	5
Standard moisture content (%)	60	60	5	4	4
Na_2SO_4 content (%)	trace	3	21	27	27
Degree of formalization (mol %)	27.5	0	0	0	0

Tables 4.22–4.24 compare the effects of various binders on sheet properties, the properties of Kuralon paper and other high-strength papers, and the prebonding strength of latex and Kuralon binder in print bonding.

VISCOSE RAYON FIBERS

Following the commercialization of its Kuralon VP fiber, Kuraray developed Eaglet S.C.R. (short-cut viscose rayon staple). This short-cut rayon is cellulosic and easily disperses in water; it can be processed on existing papermaking machinery.

S.C.R. is used in papermaking either alone or blended with other fibers, and sometimes in a combination with a web of other materials. Filters, lens polishers, paper sheets, paper clothing, tissue paper, sanitary napkins, diapers, and other products are among applications of paper made from or with S.C.R. Kuraray has been supplying various types of Eaglet S.C.R., including type PDC, which is crimped and is used for making bulky paper. According to Kuraray, in 1973 the consumption of short-cut rayon in Japan's paper production increased nearly 30% over 1972. The use of this material by Japanese papermills nearly doubled from about 6,000 MT in 1968 to about 11,500 MT in 1973, but dropped to 6,120 MT in 1974. In 1974 the exports of rayon fibers for papermaking amounted to more than 520 MT.

Some rayon manufacturers have developed special grades of rayon for various applications in papermaking, including tea-bag tissue,

Table 4.22 *Effects of Various Binders on Sheet Properties*

Binder	*Weight (g/m^2)*	*Density (g/cm^3)*	*Breaking length (km)*	*Burst*	*Stiffness by cantilever (mm)*
Kuralon VP binder fiber	44.6	0.215	3.39	325	145
Polyacrylate	46.8	0.329	2.46	546	126
Styrene-butadiene copolymer	45.1	0.313	2.61	628	136
Ethylene vinyl acetate copolymer	45.8	0.343	2.21	607	120

Source: V. M. Wolpert, "Use of Short-Cut Rayon and PVA Fiber for Papermaking Increases in Japan," *Pulp and Paper International*, 16, No. 9 (August 1974), p. 43.

Table 4.23 *Comparison of Properties between Kuralon Paper and Other High-Strength Paper*

Property	*Kuralon 100% paper*	*Kraft paper*	*Manila hemp paper*
Basic weight (g/m^2)	30.9	31.3	30.8
Density (g/cm^3)	0.34	0.60	0.62
Dry breaking length (km)	10.7	5.1	9.4
Dry elongation (%)	12.8	2.6	3.0
Net breaking length (km)	2.37	0.03	0.13
Wet elongation (%)	12.2	1.0	2.0
Tear factor	300	116	420
Burst factor	14.4	3.7	8.2
Folding endurance (times)	>50,000	43	2,300
Air permeability	0	2.0	4.0
Abrasion loss (g)	0.11	0.47	0.71

Source: Kuraray.

Table 4.24 *Difference of Prebonding Strength between Latex and Kuralon VP Binder in Print Bonding*

Composition of sheet			*Raw sheet*				*Print coated sheet*					
				Breaking length (km)					*Breaking length (km)*			
Rayon 1.5 d x 5 mm (%)	*Pulp (%)*	*Binder (%)*	*Weight (g/m^2)*	*Dry*	*Wet*	*Burst factor*	*Pick-up of latex (%)*	*Weight (g/m^2)*	*Dry*	*Wet*	*Burst factor*	*Stiffness by cantilever (mm)*
		Kuralon VP										
100	0	3	33.4	1.26	0.16	1.74	19.3	41.4	1.61	0.47	3.41	115
75	25	3	32.9	0.21	0.16	1.31	25.9	44.4	1.74	0.29	3.15	107
50	50	3	32.9	0.85	0.11	1.00	20.9	41.6	1.60	0.72	3.05	95
		Latex										
100	0	10										
75	25	10	34.5	0.21	0.07	0.96	18.2	42.2	0.73	0.19	1.94	65
50	50	10	34.2	0.47	0.23	1.02	18.6	42.0	0.83	0.30	2.07	62

Source: Kuraray.

vacuum-cleaner bag, strippable wallpaper, stencil paper, security papers (some countries with very humid climates have introduced bank notes made from pulp/rayon blends), overlay tissue for plastics laminates, etc.

Courtaulds

Courtaulds Ltd., Coventry, has developed the PM fiber belonging to the family of flat-sectioned rayons, which are made by an inflation process resulting in a hollow thin-walled fiber that subsequently collapses into a flat product. These self-bonding rayons are suitable for many applications in the paper industry, including flushable papers for sanitary products. The long fibers bond to produce a high-strength product which can be completely dispersed in a large volume of water.[31]

FMC

The American Viscose Division of FMC Corp., which has been supplying rayon to the paper industry, has developed grade RD-101, a self-bonding rayon fiber for making long-fiber speciality papers and wet-formed nonwovens. RD-101 is a short, irregularly inflated rayon of varying configuration. The bonding is achieved by the collapse of the thin walls (3 microns) which are the ribbonlike structures of the web fiber.

---

31. See Wolpert, *Synthetic and Plastics Paper*, Vol. 1, pp. 116-20.

5. Plastics Paper (extruded paper-like high density polyethylene film)

Plastics paper, or extruded paper-like HD PE film, is directed toward the packaging market, particularly for wrapping and for various types of bags for retail-shop applications, in contrast to the various types of synthetic paper which have been directed toward speciality types of printing and writing papers.[1] Compared with conventional paper, plastics paper has many superior properties, including higher tear resistance, wet strength, and grease and water resistance, as well as low gas permeability. At the same time plastics paper is price-competitive with glassine, greaseproof paper, vegetable parchment, and, to some extent, wax paper and tissue paper. Plastics paper has a lower specific weight than conventional paper, and price comparisons have to be made by area (e.g., m^2), not by weight. It has also been competing against MG kraft paper for use in bags for the retail trade.

By the end of 1975 the annual production rate of plastics paper in Western Europe amounted to about 35,000 MT. Today approximately 40 European converters are producing plastics paper. At the Plastics Industry Exhibition K-75 (Dusseldorf, October 1975) 15 extruder manu-

---

1. The definition of "plastics paper" as a paper-like HD PE film has been referred to in Chapter 2 of this book.

The author of this book has used this definition in the four-volume survey *Synthetic and Plastics Paper* and in presentations at various conferences, stressing the point that this type of HD PE film, which, on the whole, does not require any post-extrusion treatment, finds its main applications in wrapping and bag-making. The orientation of the film takes place during the extrusion process.

The definition has been accepted by a number of manufacturers and speakers at various international conferences, including J.C. Descorps and G. Stein of Sodefine, who referred to "papier plastique" at the IUPAC-EUCEPA Conference in Helsinki, June 1972. At the same conference F.J. Köhler of Reifenhäuser KG, and A. Petris of Oerlikon-Bührle AG, as well as R. Kirchhof of Paul Troester referred in German to "papierähnliche Folien," i.e. paper-like film. Italo Gigli and Renato Monzone of Rumianca at the Milano Conference, October 1972, referred to HD PE film, "now commonly called paper-like film."

Dr. W.J. Bontinck of UCB-Sidac referred at the Utrecht Conference to "plastics papers" for packaging applications.

facturers were offering machinery for making plastics paper, and orders were placed for this machinery. A very high growth rate of plastics paper production is expected.

Before the introduction of plastics paper, high-density PE film was already used for "boil-in-bag" applications, sealing media in laminates, etc. However, the introduction of this type of extruded HD PE film opened a new chapter of competition between plastics products and paper products. For the first time plastics film began to be called *plastics paper* or *paper-like plastics film*.

A few years ago, Dr. H.M. Krause of Hoechst AG (a company which has been very active in promoting the development of plastics paper in Fed Rep Germany and other countries), speaking at the demonstration of an extrusion plant, described the new products as follows:

> the term 'paper-like' is not to imply that all properties of these sheets correspond to those of paper. Such sheets rather combine the good properties of polyolefins, such as good water and gas tightness, wet strength and weldability, to a large degree with the good properties of paper, such as great stiffness, mat surface, good foldability and creaseability, and also good printability. Paper-like sheets made from HD PE will, therefore, have to be considered in future not only in competition with traditional packaging papers but also with low-density polyethylene (LD PE) sheets.

Another reference to the potential utilization of plastics paper was made by Shell Chemicals:

> The particular sector of the paper market which already is vulnerable to penetration by thermoplastics is the food and general retail packaging market in which performance advantages can be achieved by the substitution of paper by thermoplastics. The absence of acidity in high density polyethylene film has also encouraged interest in the product in certain industrial markets. In the food and general retail packaging market three product categories have been identified viz: cut sheets for hand wrapping, retail bags for point of sale use, carrier bags.[2]

Shell added that supermarket "check-out" bags may be another area of application for HD PE film. (In this application plastics paper competes not only against kraft paper but also against LD PE film, since thinner HD PE film shows the same required strength as thicker LD PE film.)

---

2. Shell Chemicals "Technical Information on Plastics: High Density Polyethylene for Paper Replacement in Packaging Applications."

JAPAN

In early 1976, thin plastics paper of 10 micron thickness and lower, made of high-molecular-weight HD PE, reached a consumption rate of over 12,000 MT/yr in Japan, and is expected to show a high growth rate in the future. Packaging of fish sold in supermarkets is an important application of this paper.

In addition to Mitsui Petrochemical, which has been pioneering special grades of HD PE polymer for this application (its Hi-Zex polymer is described in the latter part of this chapter), special grades of high-molecular-weight HD PE for making thin plastics paper have lately been developed by Nissan Chemical (grade 5001F with melt index of 0.07), Asahi Chemical, Showa Denko, and Nippon Petrochemical.

MARKET POTENTIAL

In Western Europe in 1974 the output of vegetable parchment, greaseproof paper, and glassine amounted to about 460,000 MT/yr, in North America about 280,000 MT/yr. These figures include papers made from pure chemical woodpulp and cotton fiber, and they exclude papers which have been rendered greaseproof or waterproof by coating, impregnating, or similar processes after manufacturing of the paper. These production figures indicate the potential market for plastics paper, without taking into consideration the MG sulphite wrapping-paper market (which is considerably larger than that of vegetable parchment, greaseproof paper and glassine), the penetration of which will depend on the prices of polymers and paper and the availability of these materials.

The 1972 Report of the Plastics Working Party of the Chemical E.D.C. (Britain) estimated that by 1980 more than 55,000 MT of HD PE should be used in replacement of paper, particularly tissue, greaseproof paper, and paper bags (20,000 MT of HD PE will be used for "packaging paper" and 35,000 MT of HD PE film for bags). On the basis of this forecast it can be assumed that the total West European consumption of plastics paper will amount to at least 160,000 to 180,000 MT/yr by 1980.[3]

The price increases of crude oil since the end of 1973 and the 1973–74 shortages of polymers have slowed the penetration of the market by plastics paper. During the second half of 1974 the temporary shortage of polymers was overcome. The question of how far the increase of crude oil prices will influence the competitive strength of plastics paper and other synthetics is referred to in the concluding chapter of this book. But

3. *Synthetic and Plastics Paper,* Vol. 3, p. 377.

three factors should be mentioned here:

1. The increased price of crude oil does influence the prices of finished plastics products, *but* the cost of crude oil factor gradually diminishes (is diluted) in the various subsequent processes from crude oil to monomers, polymers, films, and printed bags.
2. During the last few years the costs of converting HD PE polymers to film have been reduced by more than doubling the output rate of film on the same size extruders (in the case of high-molecular-weight HD PE on a 60-mm extruder from about 30 to over 70 kg/h). At the same time, technical progress in making bags from plastics paper has been achieved by bag-making machinery manufacturers.
3. The prices of paper increased between October 1973, and October 1974, by about 40-50%. Although the prices of paper were reduced in many countries during the following recession, the actual production costs, including high energy costs (an important factor in pulp-paper production) went up. With the end of the recession, paper prices are expected to catch up with increased production costs.[4]

A further important factor is the lower weight of plastics paper compared with that of conventional paper, which is an advantage from the point of view of the required amount of raw materials, particularly at a time when the prices of raw materials are increasing. It is estimated that the average weight replacement ratio for paper to polymer in the case of tissue-type products is about 3:1 and in the case of tough bags about 2:1. Another important advantage of bags made from plastics paper is that they occupy less room under the counter; and in department stores space is always at a premium.

PROPERTIES

There are basically two types of plastics paper, namely:

1. Plastics paper which requires higher tear strength (e.g., for bags) and is extruded from high-molecular-weight HE PE.
2. Plastics paper which does not require additional strength (e.g., for handwrapping applications) and is extruded from lower-molecular-weight HD PE.

4. According to the 1975 Annual Report of PWA, a leading West German pulp and paper manufacturer, in the Fed Rep Germany the index of paper prices (1970=100) was 107.9 in December 1973, increased to 156.3 in December 1974, and decreased to 137.7 in December 1975 and to 134.5 in March 1976. The index of woodpulp prices (1970=100) was 94.1 in December 1973, and increased to 129.2 in December 1974, 142.4 in December 1975, and 142.8 in March 1976.

The comparative properties of various kinds of plastics paper are listed on the following pages. The properties of various HD PE polymers and extruders used in the production of plastics paper are described later in this chapter.

Since plastics paper is a comparatively new product, no defined specifications of properties and quality have yet been fixed, and the properties of plastics paper supplied by individual manufacturers vary. However, it is expected that with growing consumption of plastics paper a classification of plastics paper by its technical properties, tolerances, etc., will be introduced; this will certainly be an advantage for manufacturers, converters, and users.

Comparative Properties of Plastics Paper

Table 5.1 compares the mechanical properties of HD PE tissue film with those of LD PE film, tissue paper, and greaseproof paper; and Table 5.2 compares the mechanical properties of high-molecular-weight HD PE bag film with those of 15-lb DC MG sulfite paper.

According to Hoechst AG the properties of high-molecular-weight and low-molecular-weight HD PE film, LD PE film, and standard paper compare as shown in Table 5.3

It must be added that mechanical properties of plastics paper vary depending on the production variables, including the draw ratio (resulting in orientation) of the film, as Tables 5.4 and 5.5 show.

According to Chemplex the permeability of HD PE film is lower than that of low-density and medium-density PE films (Table 5.6).

Table 5.7 shows the superior properties of plastics paper as against various types of coated paper and LD PE film, in respect to water-vapor permeability.

Coated HD PE Film

St. Regis Paper Company's Flexible Packaging Division has developed a specially coated HD PE film for the inner bags of potato chips twinpacks, replacing coated glassine paper. According to the St. Regis 1972 Annual Report, this new material extends the package's shelf life and keeps the product crisp and appetizing. St. Regis has also been supplying Pollo-Fill P, a heat-seal coated high-density PE, developed by the company's Pollock Division, and described as "a new concept in liner stock for form and fill applications."

According to the manufacturers' brochure, the MVTR (moisture vapor transmission rate) of Pollo-Fill P is more consistent than that of coated glassine. The moisture-barrier properties are built into the base

Table 5.1 *Properties of High Density and Low Density PE Films and of Selected Papers*

Property	*Test method*	*Unit*	*High density polyethylene**			*Low density polyethylene*	*Tissue paper*	*Greaseproof paper*
Weight		g/m^2	11.5	33	55	62	18	45
Thickness	BS 3422:61	micron	12.5	31.25	55.5	61.25	22.5	45
Tear strength	BS 2782:30B	MN						
MD			6.1	57.8	108	4,360	9.0	16.8
TD			1,065	1,542	675	5,960	13.9	21.7
Mullen burst strength	BS 3137:59	N/m^2	94,000	101,000	171,500	134,500	24,500	160,000
Yield tensile strength	ASTM D 882-64T	MN/m^2						
MD			21.1	31.1	28.0	12.3	39.5	97.9
TD			21.6	28.4	25.5	11.1	16.5	40.7
Breaking strength	BS 4415	N/15 mm						
MD			4.32	14.9	23.5	11.7	13.4	60.7
TD			4.41	13.6	22.4	10.5	5.6	28.1
Ultimate elongation	ASTM D 882-64T	%						
MD			390	504	709	354	1.0	1.4
TD			80	707	698	638	0.8	2.3

Note: MD = machine direction. TD = transverse direction

*Processing conditions – 60-mm extruder 200-mm die Melt temperature – 215°C Blow-up ratio – 2.5:1.

Source: Shell Chemicals, "Technical Information on Plastics: High Density Polyethylene for Paper Replacement in Packaging Applications," London.

Table 5.2 *Properties of High Density Polyethylene Bag Film and 15-lb DC MG Sulfite*

Property	*Unit*	*Test method*	*High density poly-ethylene*[1]	*High density poly-ethylene*[2]	*15-lb DC MG sulfite*
Thickness	μm	BS 3422:61	25	25	–
Tensile impact strength	Joules	Laboratory-			
MD		developed	1.0-1.8	0.9-1.6	0.19
TD		tests.	0.2-0.3	0.5-0.9	0.18
Falling dart impact strength[3]	g	ASTM 1709	70-100	70-100	20
Tear initiation	N/mm	Modified			
MD		BS 2782:30B	300-450	300-450	20
TD			500-800	500-800	25
Tear propagation	N/mm	BS 2782:30B			
MD			6-10	6-10	3
TD			30-80	30-80	4

1. High molecular weight polymer (Melt Index = 0.05) extruded at 250°C and a blow-up ratio of 4:1.
2. High molecular weight polymer (Melt Index = 0.05) extruded at 250°C and subjected to additional orientation in the tubular state.
3. Dart dropped from 14 inches.

Source: Shell Chemicals, "Technical Information on Plastics: High Density Polyethylene for Paper Replacement in Packaging Applications," London.

Table 5.3 *Properties of High- and Low-Molecular Weight HD PE Film, LD PE Film, and Standard Paper*

Property	*Unit*	*Test method*	*LD PE*	*Standard paper*	*HD PE sheet – high molecular MD/TD*	*HD PE sheet – low molecular MD/TD*
Tensile strength	kp/cm^2	DIN 53 455	150 to 200	70 to 250	450/350	330/260
Elongation break	%	DIN 53 455	600	5	650/450	500/440
Initial tear strength	P/100 μm	DIN 53 455	800	320 to 600	3,200/2,700	2,000/1,900
Impact tension strength	$kp \cdot cm/cm^2$	DIN 53 448	2,000	50 to 200	2,000/1,800	350/750
Water vapor permeability, 25 μm sheet at 20°C, 85% RH	g/m^2, 24 hours	DIN 53 122	3	extremely permeable	1 to 1.5	1 to 1.5
Oxygen permeability (aroma tightness)	$\frac{cm^2\ (NTP)}{m^2 \cdot 24\ h \cdot atm}$		6,500	extremely permeable	2,900	2,900
Maximum service temperature (short time service value in brackets)	°C		80 to 90 [95]	180 to 200	110 to 115 [120]	110 to 115 [120]

Source: Dr. H. M. Krause, Hoechst (Paper presented at the film extrusion demonstration at the Reifenhäuser extrusion plant demonstration, Troisdorf, Fed Rep Germany, February 1973). See Wolpert, *Synthetic and Plastics Paper*, Vol. 4, 1974, p. 630.

Table 5.4 *Properties of HDPE Film Produced on an 8-in Die (0.020 Die Gap) at Constant 3.3:1 Blow-up Ratio*

Nominal thickness (mils)	3	2	1.5	1	0.75	0.5
Draw ratio	2:1	3:1	4:1	6:1	8:1	12:1
Frost line height (in)	29	35	39	36	26	40
Tensile properties, MD						
Yield (psi)	4,780	4,600	4,350	4,430	4,650	4,930
Break (psi)	3,480	3,370	3,810	4,010	3,950	5,020
Elong (%)	140	180	180	202	110	350
Tensile properties, TD						
Yield (psi)	3,980	3,830	4,150	4,000	4,490	4,240
Break (psi)	3,070	4,080	3,570	4,000	4,210	3,700
Elong (%)	280	160	70	15	50	120
Tensile modulus (10^2 psi)						
MD	187	179	183	170	179	166
TD	193	188	178	167	191	178
Handle-O-Meter (rigidity), (g/in)						
MD	51.2	23.5	11.6	3.9	1.8	0.9
TD	58.6	20.8	11.8	4.3	2.3	1.2
Gauge*	2.87	2.11	1.65	1.19	0.83	0.61
TMI						
ball impact						
cm · kg	3.1	2.2	1.7	1.1	0.5	0.6
Gauge*	2.8	1.9	1.5	1.2	0.8	0.8

* Actual gauge of specimen of film tested.

Source: Chemplex Company, Rolling Meadows, Illinois, USA. See Wolpert, *Synthetic and Plastics Paper*, Vol. 1, 1971, p. 31.

Table 5.5 *Properties of HDPE Film Produced on a 3-in Die (.025 Die Gap)*

Nominal thickness (mils)	2	1.3	0.5
Blow ratio	4:1	5:1	5.2:1
Draw ratio	3.1:1	3.9:1	9.6:1
Frost line height (in)	9½	8½	10
Tensile properties, MD			
Yield (psi)	4,440	4,590	4,980
Break (psi)	3,850	4,220	6,350
Elong (%)	350	320	440
Tensile properties, TD			
Yield (psi)	4,500	4,600	4,770
Break (psi)	3,060	3,940	5,120
Elong (%)	320	320	580
Tensile modulus (10^3 psi)			
MD	181	186	177
TD	195	192	204
Handle-O-Meter (rigidity), (g/in)			
MD	20.6	5.6	0.7
TD	18.9	6.6	0.9
Gauge*	2.18	1.26	0.59
TMI			
ball impact			
cm · kg	2.9	1.8	0.9
Gauge*	2.2	1.2	0.5

* Actual gauge of sample of film tested.

Source: Chemplex Company, Rolling Meadows, Illinois. See Wolpert, *Synthetic and Plastics Paper*, Vol. 1, 1971, p. 31.

Table 5.6 *Typical Values for Permeability of Polyethylene Films*

	LD PE	*MD PE*	*HD PE*
MVTR	17	12	7
CO_2	40,000	10,400	7,000
O_2	6,600	4,700	2,100
N_2	2,800	1,700	650

Notes: Values based on film thickness of 1 mil (.001").

To obtain values for 100 in^2 divide by 15.5.

Moisture vapor transmission rate = g/m^2/24 hr, 100° F, 90 to 95% RH.

Gas = cc/m^2/24 hr, 1 atm.

Source: Chemplex Company, Rolling Meadows, Illinois. See Wolpert, *Synthetic and Plastics Paper*, Vol. 1, 1971, p. 29.

Table 5.7 *Comparative Properties of Packaging Coatings*

Material and weight (lb/3,000 ft²)	*(in²/lb)*	*Water vapor permeability (g/m²/24 hr), 90% RH, 100° F*
Wax coatings		
Machine finished sulfite 25 lb, refined paraffin 5 lb	14,380	High*
Supercalendered sulfite 25 lb, refined paraffin 10 lb	12,300	24**
Glassine 25 lb, refined paraffin 6 lb	13,900	4**
Supercalendered sulfite 25 lb, Hot Melt 10 lb	12,300	High*
Solvent coatings		
Supercalendered kraft 32 lb, cellulose nitrate 8 lb	10,800	25
Glassine 22 lb, cellulose nitrate 3 lb	17,250	10
Emulsion coatings		
Supercalendered kraft 25 lb, vinylidene chloride (Saran) 6 lb	13,900	6.5
Glassine 25 lb, vinylidene chloride (Saran) 4 lb	14,900	6.0
Extruded coatings		
Supercalendered kraft 25 lb, LD PE (1 mil) 15 lb	10,800	21
Supercalendered kraft 30 lb, HD PE (1 mil) 15 lb	9,600	11
Blown film		
Low density polyethylene (1 mil) 14.5 lb,	30,000	17
Sclair 19A film (1 mil) 15 lb	29,000	4.7

* The reading was too high to be accurately tabulated.

** Once creased, the coating becomes almost ineffective.

Source: Du Pont of Canada Ltd. See Wolpert, *Synthetic and Plastics Paper*, Vol. 1, 1971, p. 37.

Table 5.8 *Comparison of Physical Properties of Pollo-Fill P and Some Coated Materials*

Material	*Standard*	*2 mil HD PE/ 2.5 lb/ream Heat seal coating*	*2 lb EVA/ 28 lb Glassine & 4 lb PVDC*	*3 lb EVA/ 28 lb Glassine & 3 lb EVA*	*140 Gauge polymer coated cellophane*
Basic weight	lb/ream	32.0	34.0	34.0	31.0
Yield	in^2/lb	13,500	12,700	12,700	14,000
Durability case drop	2-4 foot case drops at 0°C	excellent	fair	fair	fair
Durability freezer tumble	Tumbled at 5 RPM for 10 minutes at 0°F	excellent	poor	poor	poor
MVTR-Cup method	24 hr at 100°F and 90% RH, flat/creased	0.22/0.23	0.24/0.26	0.26/0.28	0.70/0.77
Tear	MD/TD	48/192	34/32	32/32	40/49
Stiffness	MD/TD	1,800/1,400	2,650/1,700	2,350/1,650	1,450/750

Source: Wolpert, *Synthetic and Plastics Paper*, Vol. 4, 1974, p. 639, based on information supplied by St. Regis Company's Flexible Packaging Division (previously known as Pollock Division).

sheet rather than into heat-seal barrier coatings. As a result, the barrier properties do not deteriorate either in machining or handling, as is the case with materials which have the barrier properties built into the heat-seal coatings. Pollo-Fill P contains no moisture. This eliminates the problems of drying out and becoming brittle.

It is claimed that creasing, heat-sealing, and humidity conditions do not affect the grease-barrier properties, and that these properties are therefore superior to those of coated glassine. Table 5.8 compares the physical properties of Pollo-Fill P with those of some other coated materials.

Plastics Paper from HD PE/Elastomer Blends

Paper-like films extruded from HD PE/Kaolin and HD PE/styrene compounds were described in Chapter 3 since these films aim mainly at the printing paper markets. The paper-like films extruded from HD PE/elastomers compounds are described in the following pages since it is expected that the main potential application of this film will be in the packaging field.

HD PE/Elastomer Blends

Essochem Research, Diegem (Belgium) has developed methods of extruding paper-like film from HD PE/elastomer compounds.[5] It is felt that although elastomers are more expensive on weight basis than HD PE, and although the compounding of HD PE with elastomers represents an additional cost factor, there are certain advantages of using these compounds because of their processability characteristics, the lower specific gravity of elastomers, and the properties of films made from these compounds, all of which outweighs the additional cost of raw material. These compounds produce tougher and stronger films.

Various elastomers can be used for this film production, including Vistanex L-100, an Esso polyisobutylene of 0.92 specific gravity and Vistalon 5600, an Esso EPDM (ethylene propylene terpolymer) of 0.86 specific gravity.[6] Films of lower thicknesses made from these compounds have the same strength properties as corresponding thicker films made exclusively from HD PE. The choice of which elastomer to use in the compound depends on the end product. A big advantage of these compounds is the higher output of extruded film per hour, and the possibility of using extrusion equipment that is cheaper than that used for HD PE film. LD PE extrusion lines, including adiabatic extruders, can be used in the production of these films. The bubble stability of films made from these compounds is higher than that of HD PE films. Tables 5.9 and 5.10 list the tear properties of these films.

It is expected that these types of film will find application not only in packaging but also in other fields. Trials of making IBM tape from this film are reported to be in progress. (HD PE films have been already used for making one-way typewritter ribbons.) These films are claimed to have good printing characteristics, with a "bare minimum of discharged treatment" required. A slightly higher slip-additives content is required to compensate for the effect of elastomers.

Table 5.9 shows that the 20-micron paper-like HD PE film containing Vistanex L-100 registers significant increases of dart drop impact and Elmendorf tear strength compared with 100% HD PE film. In this thickness, film made from this compound is more suitable for paper replacement than the film made from a compound containing Vistalon 5600, which is less paper-like and of somewhat poorer color, but which has a very high tear strength. Films of higher thicknesses (e.g., for

5. Essochem Research Center, Diegem, Belgium, "Elastomer-HD Polyethylene Blends for Paper-like Thin Film," September 1972; "High Strength HD PE Thin Film," November 1973; "High Impact and Tear Resistant HD PE Films," November 1973.

6. Esso Chemical (USA) "Specifications: Esso Chemical synthetic rubbers," and "Vistalon 5600," Document RBE.46SU.73, 1973.

Table 5.9 *High Strength HD PE Thin Film*

	Control	1	2	3	4
Formulations					
HD PE (MFI 0.2)	100	90	80	90	80
Vistanex L-100	–	10	20	–	–
Vistalon 5600	–	–	–	10	20
BHT	–	0.07	0.07	0.07	0.07
Properties					
MFI (g/10 min)					
190°C, 2.16 kg	0.48	0.40	0.31	0.35	0.29
Dart drop impact (g/mil)	58.1	120.7	170.5	73.7	122.2
Elmendorf tear (g/mil)					
MD	17.9	27.1	54.8	35.0	161.8
TD	151.9	251.2	290.3	323.6	479.5
Color	——— white ———			—— straw ——	

Source: Essochem Research Center, "High Strength HD PE Thin Film," Diegem, Belgium, November 1973.

Table 5.10 *High Impact and Tear Resistant HD PE Films*

	Control		3		4	
Formulations						
HD PE (MFI 0.3)	100		80		80	
Vistanex L-100	–		20		–	
Vistalon 5600	–		–		20	
BHT	–		0.07		0.07	
Properties						
Film thickness (micron)	50	80	50	80	50	80
Dart drop impact (g/mil)	34.5	44.9	116.3	134.4	64.9	60.2
Elmendorf tear (g/mil)						
MD	16.4	32.0	31.1	53.4	53.9	93.0
TD	27.7	25.2	106.1	66.9	139.4	131.9

Source: Essochem Research Center, "High Impact and Tear Resistant HD PE Films," Diegem, Belgium, November 1973. See Wolpert, *Synthetic and Plastics Paper*, Vol. 4, 1974, p. 654.

Table 5.11 *Adiabatic Extruder Trial Results (Compound – 15% Vistanex L-100 in Rigidex 9 HD PE)*

Flat width (cm)	BUR	Thickness (micron)	Freeze line (in/cm)	Elmendorf tear MD (g/mil)	Elmendorf tear TD (g/mil)	Dart drop (g/mil)
Group I (150 mm die)						
53.0	2.25	23	constant	18.2	753	39.6
64.0	2.74	21		22.1	806	50.2
79.0	3.35	22	(unknown)	21.3	661	51.9
94.5	4.0	23		22.4	392	39.1
51.7	2.25 "average"	130	constant	74.9	92	>500
53.0		66		47.1	127	>300
53.5		41	(unknown)	28.2	264	117.8
53.0		23		18.2	753	39.6
50.5	2.25 "average"	45	11/31	26.1	596	98.7
52.7		41	22/62	26.6	310	104.2
55.5		40	34/86.4	26.5	143	85.7
Group II (100 mm die)						
67	4.2 "average"	36	24/67	28.8	234	132.7
64		39	15/38.1	25.5	325	130.2
62		58	15/38.1	38.6	108	157.7
65		56	25/69.6	46.4	90	158.9
68		55	33/84	50.3	67	158.0
71		36	33/84	31.4	180	129.9
71		27	33/84	17.1	657.8	64.9

Source: Essochem Research Center, "Elastomer – HD Polyethylene Blends for Paper-like Thin Film," Diegem, Belgium, September 1972. See Wolpert, *Synthetic and Plastics Paper*, Vol. 4, 1974, p. 655.

making carrier bags and for building-trade applications) made from HD PE/Vistalon 5600 compounds have the required characteristics. These films of 50 and 80 microns were successfully extruded at a blow-up ratio of 3:1 on low-density polyethylene film extruders. Vistalon blend uses more power than Vistanex blend and generates more neck pressure.

In the production of paper-like film of about 20 microns, trials with 15% Vistanex L-100 and 85% Rigidex 9 (an HD PE polymer manufactured by BP Chemicals) compound, using an adiabatic LD PE blown-film extruder with a 150-mm die, have registered the optimum impact strength at a blow-up ratio of about 3.5:1, and a 70- to 80-cm freeze-line height. In the case of thicker films, a lower blow-up ratio produced satisfactory results; Tables 5.11 and 5.12 show trial results.

Table 5.12 *Adiabatic Film Extrusion*

Compound	*1*		*2*
HD PE	85		80
Vistanex L100	15		20
BHT	0.07		0.07
Sample no.	*A*	*B*	*C*
Blow-up ratio	3.35	3.25	4.1 plus
Thickness (microns)	18	18	12
Properties			
Dart drop (g/mil)	27.8	79.5	170.0
Tear (g/mil)			
MD	19.0	20.1	24.0
TD	468	500	583

(Extruder – 40 mm)

Temperature (°C)	175-190-200
Screw speed (rpm)	322
Amps	18
Die	150 mm

Source: Essochem Research Center, "Elastomer – HD Polyethylene Blends for Paper-like Thin Film," Diegem, Belgium, September 1972. See Wolpert, *Synthetic and Plastics Paper*, Vol. 4, 1974, p. 655.

MARKETS AND MANUFACTURERS

The three main applications of plastics paper, by thickness, are:

1. 6-12 microns (12 microns, about ½ mil) replacing tissue for wrapping, with wet-wrap and greaseproof properties;
2. 15-35 microns (25 microns, about 1 mil) tough, nonsplitting film for retail bags;
3. 35-75 microns (72 microns, about 3 mils) for carrier and grocery bags.

The market penetration by plastics paper varies in different countries, but there is an overall trend of growing consumption. The commercialized use of plastics paper, including applications in laminates and applications outside the markets just mentioned, are referred to in the following pages.

At the end of 1975 and the beginning of 1976 European production of plastics paper (paper-like HD PE film) was at a rate of about 35,000

MT/yr, and it was forecast that by 1980–81 the demand for plastics paper would amount to about 70,000 MT/yr.[7]

At the TAPPI 1976 International Symposium A. Petris of Oerlikon-Buehrle described the actual and potential markets of plastics paper in Europe, and said that the most important market for plastics paper is bags made from plastics paper of 15–30 micron thickness. The bags are supplied as flat bags, blocked tear-off bags, side-gusseted bags, and serve as packaging material for food, textiles, haberdashery, stationery, drugs, toilet articles and tobacco products. In households this plastics paper is also used for preservation and refrigeration purposes, as well as for welded bags for sealing medical instruments. Thicker bags are mainly used as cooking bags and for storing deep-freeze products.

According to Petris, the second biggest market is in wrapping applications, replacing tissue paper. Plastics paper has a high degree of impermeability, and is used for wrapping cut meat and sausages, fish, dairy products, flowers, etc. Petris said he knew customers who were supplying plastics paper to wholesale butchers for wrapping large chunks of meat. Carrier bags are the third biggest market, and this market has been growing steadily, since free-standing HD PE bags facilitate the filling of these bags at the check-counters. Flat-welded bags and carrier bags have been used in textile shops, chemist shops, book stores, and stationery department stores. He also referred to the potential use of plastics paper as laminating material (e.g., as a covering film for self-sealing labels, posters, and visiting cards).

Petris said that interesting growth rates can be anticipated in the three main markets. He referred to Swiss regulations limiting the weight of wrapping materials to 3% of the weight of wrapped food. In butcher shops the wrapping is weighed together with the meat purchased, and the Swiss Consumer Forum demands that butchers use lightweight, leakproof wrapping materials. Plastics paper has the advantage of a lower weight than conventional wrapping paper, and meets the requirement of being leakproof.

United Kingdom

Among European countries the UK has been leading in the consumption of plastics paper. A number of leading department stores, including Harrods, Selfridges, and Marks & Spencer (who have a chain of food and textile shops), have been using bags made from plastics paper. In butcher shops, fish shops, and flower shops, plastics paper has been used in hand-wrapping applications.

---

7. A Petris, "Paper-like Film" (Presentation at the TAPPI Synthetic Papers & Pulps International Symposium, Washington, D.C., March 31-April 2, 1976).

The number of British paper manufacturers who have entered the field of production of plastics paper has been steadily increasing. Sir Don Ryder, then Chairman of Reed International, a leading British papermaking concern, in his statement to the Company's Annual Meeting (August 2, 1973), said that the subsidiary of Reed Medway Sacks, The Medway Plastics, "which produces high density polyethylene film and bags, has also made considerable progress during the year, and now occupies a leading place in *this important and growing market* [italics mine]."

Among prominent British paper and plastics film converters, Bibby & Baron, which started converting HD PE film to department store bags in 1972, has in the meantime installed extrusion equipment for HD PE film for bag conversion and printing at its Bury factory. The high-molecular-weight HD PE film is supplied in a range of 15- to 24-micron thicknesses, with a recommended 18-micron minimum thickness for grocery- and department-store bags. These bags are flexographic printed and can be supplied in up to four colors on one side or in two colors on both sides. Venus Packaging supplies plastics paper for the shoe retail trade, including shoe-cartons (14-micron thickness) and printed shoe-bags. C.K. Addison Co. Ltd. is among the first and biggest plastics paper manufacturers.

Industrial Applications

Within the electronics industry, Honeywell uses plastics paper to package instruments to prevent dust particles and water penetration (replacing previously used acid-free paper). Cosmetics manufacturers (e.g., Innoxa England Ltd.) use plastics paper for lining trays in which liquid lipstick is poured. The plastics paper keeps the lipstick moist for a lengthy period. Max Factor packs high-gloss containers in plastics paper to prevent scuffing. Among food manufacturers, G. Brazil & Co. has been wrapping uncooked boned hams in this film. The wrapped ham is being cooked in pots for up to nine hours. Henry Telfer & Co. uses plastics paper in its meat-pie production line to reduce hot-fat losses and to avoid shortcomings which occurred when using greaseproof paper.

Fed Rep Germany

In Fed Rep Germany, where plastics paper has been called HM-Folie (high-molecular-weight film), it has generally been realized that for some applications (e.g., handwrapping of meat, fish, flowers, etc., sold in retail shops) film of sufficient strength for these applications is more economically produced from lower-molecular-weight polymers; this realization has resulted in the introduction of MM-Folie (medium-molecular-weight film).

The penetration of the market by plastics paper in Fed Rep Germany has been slower than in Britain, partly, it is claimed, as a result of paper manufacturers fending-off the plastics paper competition by selling conventional paper at very low, uneconomical prices. Also, only smaller and medium converters were active in the production of plastics paper. The situation changed in 1974 when, for example, bigger converters installed plants for making plastics paper. Melitta-Werke, Bentz & Sohn, Minden, well-known manufacturer of coffee filters and allied products, which started some time ago marketing sandwich bags made from about 20-micron plastics paper (replacing imitation parchment paper), is reported to have installed extrusion equipment for manufacturing HD PE film. Folie Forchheim GmbH, Forchheim, supplies plastics paper for pre-cooked and cooked food, since the low air permeability of this paper enables lengthy storage of food. This paper can be converted on bag form-fill and seal machines. Forchheim and other German manufacturers find interesting applications for HD PE films, including those of greater thicknesses, for laminates with paper, other plastics films, and aluminum foil for packaging and technical purposes.

HD PE sheets/polyester film laminates are used for packaging food, including individual-portion packs by leading German food manufacturers. Herta KG, a leading German food manufacturer, uses laminates of 75-micron HD PE/12-micron polyester film for food packs in sizeable quantities, applying Multivac-R70 machines built by Multivac KG of Germany. In some cases meat portions in vacuum-formed packs are even made from considerably thicker (e.g., 400 micron) HD PE sheets/15-micron polyester film. The supplier of these laminates is Kalle, Wiesbaden-Bieberich.

The number of department stores using flat retail bags made from plastics paper has been increasing, and in 1974 these bags were also introduced in some Woolworth shops in Germany. During the shortage of polymers which occurred in 1973 and the first half of 1974, converters were able on some occasions to secure HD PE lots instead of the usual LD PE material, and reported good experience with it; the clients were satisfied, and they have now introduced a regular production of plastics paper.

Additional extruders for making plastics paper were installed in 1975, and orders for extruders for 1976 delivery were placed with extruder manufacturers. In early 1976 there were 14 manufacturers of plastics paper in Fed Rep Germany, and the production of plastics paper was expected to show a high growth rate, at least during the next five years.

Italy

Plastics paper has been introduced for wrapping butter (0.5-kg packs).

France

Forty-micron plastics paper coated with PVdC has been used for packing coffee (1-kg packs). In this application the coated plastics paper replaces aluminum foil, and is claimed that in plastics paper packs a perfect aroma retention of coffee has been achieved.

Finland

UPO Oy has developed plastics paper for packaging applications as well as for making geographical maps, and has also been offering laminates of HD PE film with various materials, including one with aluminum foil.

Austria

Plastics paper of about 15-micron thickness has been used for wrapping carcasses for the supply to butchers. In this particular application the electrostatic properties of the film help to make the film cling (wrap tightly) to the carcass.

Belgium

Printed plastics paper of 40-micron thickness, made by Unisac, Tournai, is used as gift-wrapping material.

Ireland

Black refuse sacks made from HD PE film have recently been manufactured and exported by Paclene Company Ltd., Ennis, Co. Clare, replacing refuse sacks made from paper and also from LD PE film (HD PE film sacks of lower gauges have the same strength as those made from thicker LD PE film). Eighteen-micron plastics paper is now being used for packing individual lettuces.

Plastics paper has a high thermal stability, which enables this material to be used for packaging asbestos mats at 100°C, and as a release film for rubber compounds. According to some reports, plastics paper has been used for covering the sprayed motorcar bodywork; here, as in packaging meat carcasses, the electrostatic properties of this material are an advantage.

Australia

Jiropak Pty. has developed a packaging system for tetrahedral packs for liquid milk and other fluids by using HD PE film.

North America

The introduction of paper-like HD PE film for wrapping and small bags used by department stores and butchers has been slower in North America than in Western Europe. Some manufacturers feel that one of the reasons for the slow penetration of these markets in North America has been the low-profit margin for converters and the fact that special equipment is required. Nevertheless, HD PE film, either coated with a heat-seal coating or coextruded with EVA, has secured a market in North America for potato chips and other snack food applications, replacing glassine structures and PVdC composites. HD PE films for this application are either white pigmented or natural colored, and their thickness is in the range of 1¼ to 2 mils (30–48 microns).

Valeron, a special laminate of two oriented high molecular weight HD PE sheets, competing with 5- and 6-ply kraft paper, was described in Chapter 1 of this book.

Grocery Bags

It is estimated that the U.S. grocery-bag market is about 12,000 million bags per year, and it is believed that 3-mil HD PE film bags have a good potential of penetrating this market.[8] The system of producing freestanding, self-opening-square, flat-bottom grocery bags from high-density polyethylene was developed by National Petro Chemicals Corporation, New York (a joint venture of National Distillers and Chemical Corp. [NDCC] and Owens-Illinois); U.S. Industrial Chemicals Co. (a division of NDCC), which supplies the speciality grades of HD PE; and H.G. Weber & Co. Inc., Kiel, Wisconsin, which supplies the bag-making equipment.

A pilot plant (from resin to finished printed sack) has been in operation at the U.S.I. works at Tuscola, Illinois, and National Petro Chemicals offers licenses for this system. The plant consists of film-extruding and

8. *Synthetic and Plastics Paper*, Vol. 4, pp. 640-43, based on information supplied to author by National Petro Chemicals Corporation, New York, 1973 and 1974.

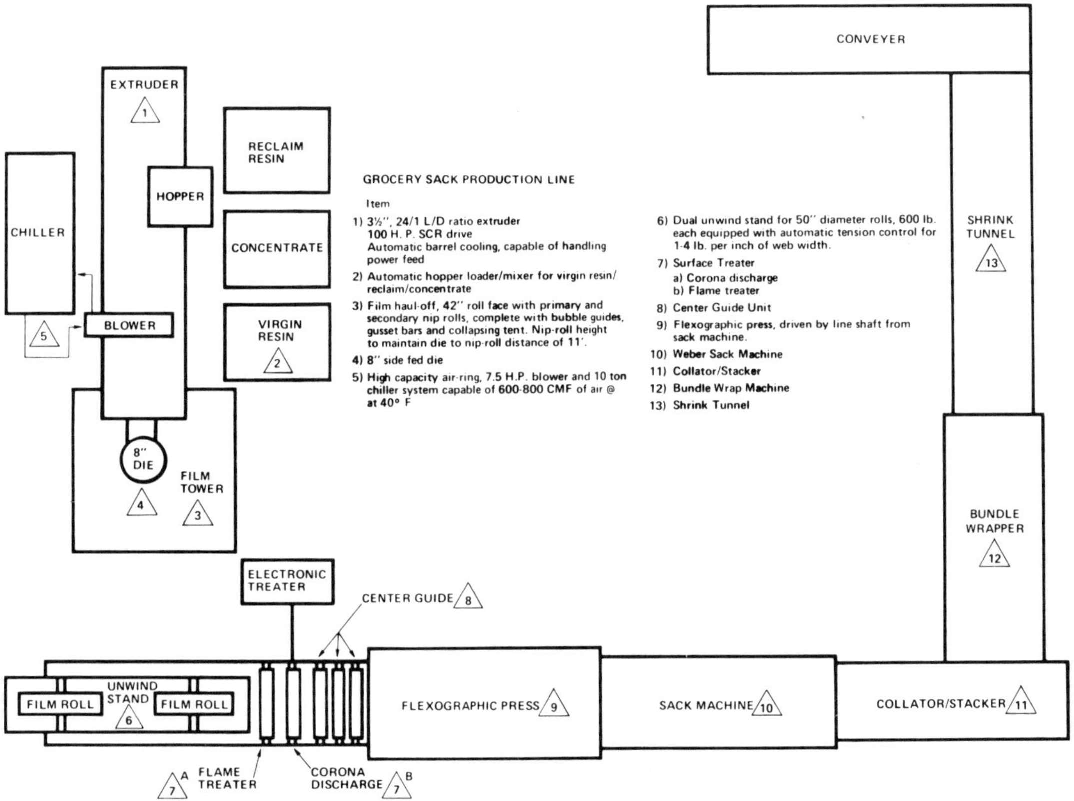

Diagram of HD PE grocery sack production line. Courtesy National Petro Chemicals Corporation.

bag-making equipment, and has a production capacity of 120 1/6-bbl HD PE film bags of about 3-mil thickness per minute (min).[9] The tubing is made on a 3½-in. film extruder, then printed, heat-sealed, cut to sack length, folded with a flat bottom, and stacked and wrapped in shrink film, all in one operation. IGA (Independent Grocers' Alliance) was the company's first client. IGA is reported to have a nationwide network of distributors serving about 3,700 grocers, some of whom have started replacing kraft paper sacks with the plastics paper sacks made at the Tuscola plant. The sacks are white with the IGA emblem in red. They are sold under the trade name Plasti-Tote and have the following imprint:

> IGA's new translucent plastic grocery bag tough, sanitary, moisture proof, light weight, re-usable.
> NPCC U.S. PAT. NOS. 3.606.822 & FOREIGN PAT. PEND.
> This polyethylene grocery sack is not detrimental to the environment in its manufacture, use or disposal. As a waste container it will help to reduce odor, pest infestation and litter. When burned only harmless carbon dioxide, water vapor, nitrogen and oxygen gases are given off.

According to IGA, Plasti-Tote sacks have many advantages over kraft paper sacks, since they require 60% less storage space and are 50% lighter. In addition, they are stronger than 80-lb or duplex 40/40 kraft paper sacks. The Plasti-Tote bag is not weakened by moisture and can be re-used at home to store food (even frozen food) or to line the garbage can.

H.G. Weber (USA) developed the Poly Sack Machine, which because of its high operating speeds does not run in-line with extruders. The machine receives wound rolls of pregusseted tube stock and produces the bags at speeds of up to 120 bags/min (i.e., 240 ft/min), since the seal cut-off length of the bag is 21 in. However, further developments of Weber's equipment have resulted in higher-speed operation. Flexographic equipment within the range of the Prestoflex printing presses is also supplied by Weber.

According to National Petro Chemicals the production costs of pigmented and printed sacks are approximately $19.50/1,000 sacks, based on the price of HD PE film-grade resin at 14¢/lb. National Petro Chemicals estimates that a single-line system, capable of producing 84 million sacks/yr, would require an investment of approximately $1.2 million, covering land, building, rail-siding investment, working capital, one bag-making machine with related equipment, and five 3½-in. extruders to keep the system operating around the clock.

---

9. Ibid.

Plasti-Tote HD PE grocery sack production machine. Courtesy National Petro Chemicals Corporation.

It is claimed that HD PE bags, because of their advantages over kraft paper bags, are bound to secure a growing share of the market despite the price difference. National Petro Chemicals feels that when the price difference decreases to 15% the HD PE bags could secure a 20% share of the total market, that is, 2,400 million bags/yr, requiring over 80,000 tons of HD PE resin for this application. Although National Petro Chemicals Corporation has secured one licensee in the USA for production of these HD PE sacks, the shortage of HD PE and the price increases of polymers in 1974 retarded the large-scale market penetration by these sacks, despite their functional advantages, and may temporarily limit their use to speciality applications.

EMBOSSED PLASTICS PAPER

Buyers for some department stores, which are potential large-scale users of plastics paper for hand-wrapping applications, have expressed the opinion that they would introduce this material in their stores, if it were supplied finely embossed.

Chemplex Company (USA), manufacturer of HD PE polymers, feels that embossed films show an increased strength and increased resistance to tearing or splitting. In an embossed film, less resin can be used (the gauge can be thinner) and still the bulk is increased. Printing can be done by the gravure process, and embossing can be done by the cold process (rubber against steel embossing roll).

Investigations by Chemplex have shown that the splittiness (low machine direction tear strength) of blown HD PE film may be reduced by embossing it with a pattern consisting of a series of plain or interrupted fine horizontal lines parallel to the transverse direction (i.e., across the machine direction of the film). It is believed that this increase in MD tear strength is caused not only by the physical interruption of the tear by the embossed ridges in the film, but also by a beneficial change in the crystalline orientation of the film in the embossed area, which also adds to its tear strength. Thus, by embossing the film with an appropriate pattern, several or all of the following benefits may be obtained.

1. A lower blow-up ratio can be used for a given web width permitting use of a larger die with increased extrusion output and reduced extrusion costs.
2. The increased strength may allow a reduction in gauge, saving material and reducing costs.
3. Embossing increases the bulk of the product and its apparent stiff-

ness. This in many cases complements the second benefit.
4. Appearance of the product is enhanced.[10]

H.-J. Ploenes, of Hoechst AG, speaking at the Reifenhäuser extrusion plant demonstration (February 1973), felt that embossing imparts to paper-like HD PE film a bulkier character, an excellent handle, and a marked increase in stiffness. This latter property is desirable, particularly for very thin film between 10 and 20 microns (e.g., for wrapping applications).

Paper-like plastics film can be perfectly embossed on the calenders normally used in the paper industry. Films below a thickness of about 30 microns can be cold embossed, both with coarse and fine patterns, on an embossing calender equipped with a steel and vulcanite roller. When the patterns are very fine, heat embossing should be employed at temperatures around 70°C in order to obtain high sharpness of the contour. Thicker films (e.g., between 50 and 100 microns) can be readily cold embossed, if the pattern is relatively coarse. The finer the pattern, the sooner heat embossing is necessary. However, very fine patterns, even at highly elevated temperatures, do not produce sharp contours. Paper-coated repeat rollers can be used, but they improve contour sharpness only to a small degree.[11]

HD PE POLYMERS FOR PLASTICS PAPER

Leading manufacturers of HD PE polymers have developed special grades of homopolymers and copolymers for making different types of plastics paper. Some manufacturers feel that homopolymers are to be recommended when rigidity and softening point are of main importance, whereas copolymers are to be recommended when greater flexibility, maximum resistance to cracking, and better long-term load-bearing properties are required.[12]

The properties of the individual polymer grade used for making plastics paper are important in the quality and output of the film to be produced. The polymer grade must be chosen in relation to the type of plastics paper to be made, and it is felt that a good-quality extruder can be adapted to whatever polymer grade is chosen. Because polymer-grade properties are so important, polymer manufacturers have con-

10. *Synthetic and Plastics Paper*, Vol. 2, p. 200, information supplied to author by Chemplex, 1971.

11. Hans-Joseph Ploenes' presentation at the Reifenhäuser extrusion plant demonstration, Troisdorf, Fed Rep Germany, February 1973.

12. *Synthetic and Plastics Paper*, Vol. 1, p. 17, based on information supplied by BP Chemical (UK) Ltd.

ducted continuous work on the development and improvement of their special grades of HD PE for various types of plastics paper.

HD PE polymers are supplied in a range of densities and melt-index properties. Details of properties of various HD PE polymers, supplied by some leading manufacturers, are listed in the following pages in order to indicate the range of available HD PE grades for making plastics paper.

High-density PE has a higher degree of crystallinity than low-density PE, is produced by low-pressure processes, and is basically linear in structure (with a small number of short side branches).

HD PE film is stiffer than LD PE film and has a less-waxy feel (handle). According to Hoechst AG, the HD PE polymer Hostalen G has a crystallinity figure of 60–80% at a density of 0.942–0.965 g/cm³. The density value is directly related to the degree of crystallinity, and with increased density (crystallinity) an increase also occurs in the following properties:

1. Crystalline melting range.
2. Yield stress (tensile strength), rigidity.
3. Hardness and modulus of elasticity and/or torsional rigidity.
4. Resistance to solvent attack.
5. Impermeability to gases and vapors.[13]

According to Du Pont of Canada, grade Sclair 17A HD PE for making plastics paper is composed of long molecular chains with comparatively few short side branches, and it is highly crystalline in its solid state.

The molecular weight of HD PE polymers is an important property; the higher the molecular weight, the greater the number of points of attraction and entanglement among molecules. In fact, the introduction of high-molecular-weight HD PE has enabled the making of paper-like films with greater tensile strength, which is a prerequisite for many applications in bag making, as already mentioned.

According to Shell Chemical, high density PE is highly crystalline in the solid state, resulting in the blown film becoming opaque when the cooling of the "web" is done slowly.[14] If the film web is cooled rapidly, as is the case in the cast-film extrusion, a clearer product is obtained. In addition to opacity, the crystallinity of HD PE (as indicated by density) is important in determining other properties. Thus, increasing the density gives increased yield tensile strength and stiffness (crispness), and

13. Hoechst AG "Hoechst Plastics: Hostalen G, High Density Polyethylene," Bulletin, May 1969.

14. Shell Chemicals, "Technical Information on Plastics: High Density Polyethylene for Paper Replacement in Packaging Applications," London.

better barrier properties. While the highest-density polymers appear to be the most suitable from the point of view of application, orientation sensitivity (tendency to split) and film shrinkage on cooling (creasing) become more critical when resin density is increased.

Melt-Flow Index

Molecular weight, measured by the melt-flow index, has an important effect on the film properties and the processing operation. The lower the polymer melt index, the tougher the film produced and the better the bubble stability. However, extruder output is reduced for the low-melt-index polymers.

In the following pages regarding various grades of HD PE, reference is made to the melt index (melt-flow index) of these grades. The melt index is a measure of the melt viscosity of the polymer and is an important parameter to the film maker. The melt-flow index is measured by the weight of the material (expressed in grams) extruded in 10 minutes under a constant deadweight and at a constant temperature.

A low melt index corresponds to a high melt viscosity. The relationship between melt viscosity and the melt-flow index is a logarithmic one. In the extrusion process, high melt viscosity (low melt flow) and low melt-flow index grades of polymers are used.[15, 16] In some cases the MI (melt index based on 2,160 kg) is expressed in HLMI (based on 21,600 kg). Sodefine states that its high-molecular-weight HD PE grade 4710 has an MI of 0.05 and an HLMI of 8.0.

Individual HD PE Polymers

Hi-Zex

This polymer, developed and manufactured by Mitsui Petrochemical Industries Ltd. (Japan) using the Ziegler process, includes grade 7000F for the production of tissue-type plastics paper. Grade 7000F has the properties shown in Table 5.13. Because of its linear molecular structure, Grade 7000F has the characteristics of low dielectric loss and high insulation resistance. The main applications of this grade are in the wrapping and packaging field. According to some plastics paper manufacturers, Hi-Zex has very good processibility characteristics.

15. A.J. Brydson, *Plastics Materials* (London: Iliffe Books Ltd., and Princeton, New Jersey, USA: D. Van Nostrand Company, 1966).

16. J.H. Briston and L.L Katan, "Plastics Films," (published for The Plastics Institute by Iliffe Books, London, 1974, The Butterworth Group).

Table 5.13 *Properties of Hi-Zex Grade 7000 F*

Property	*Unit*	*Test method*	*Value*
Average molecular weight		Viscosity method	12.0×10^4
Melt flow index	g/10 mm	ASTM D1238-65T	0.04
Density	g/cc	ASTM D1248-65T	0.955
		ASTM D1505-63T	0.955
Tensile yield stress	kg/cm^2	ASTM D638-64T	260
Elongation at yield	%	ASTM D638-64T	20
Tensile strength	kg/cm^2	ASTM D638-64T	350
Ultimate elongation	%	ASTM D638-64T	900
Izod impact strength (notched)	$kg \cdot cm/cm^2$	ASTM D256-56	25
Stiffness in flexure	kg/cm^2	ASTM D747-63	9000
Hardness	Rockwell R	ASTM D785-65	55
Vicat softening point	°C	ASTM D1525-65T	130
Melting point	°C	ASTM D2117-64	132
Brittleness temperature	°C	ASTM D746-64T	−130

Source: Wolpert, *Synthetic and Plastics Paper*, Vol. 1, 1971, p.38, based on information supplied by Mitsui Petrochemical Industries.

Rigidex Type 9

This polymer is manufactured by BP Chemicals Ltd.(UK), using the Phillips solution process. It is used for making tissue-type plastics paper. Rigidex 9 has a melt index of 0.9, and has the physical properties shown in Table 5.14.

In addition to the homopolymer type 9, BP Chemicals has also developed several types of copolymers which find application in making blown film including type 2000 with a density of 0.954 and the low melt index (g/10 min) of 0.2. It has a softening point (Vicat) at 125°C.

Hostalen G

This HD PE polymer is manufactured by Hoechst AG (Fed Rep Germany), which has been very active in the promotion of paper-like HD PE films. Hoechst has been supplying various grades of Hostalen G for paper-like film extrusion, and has lately been concentrating on high-molecular-weight grade GM 9955 F2 supplied in granules and on grade LP 103 supplied in powder. Granules incorporating antistatic agents can be supplied; in the case of powder, however, the antistatic agents are, if necessary, added by the film manufacturer. The properties of these polymers are shown in Table 5.15.

Table 5.14 *Properties of Rigidex 9*

Property	*Unit*	*Value*	*Unit*	*Value*
Density	g/ml	0.96	kg/m^3	960
Softening point (Vicat), (°C)		128		
Tensile strength	lb/in^2	4,200	kg/cm^2	295
Elongation at break (%)		>300		
Flexural modulus	lb/in^2	2×10^5	kg/cm^2	14×10^3
Cold bend (°C)		<−65		
Cold shatter (°C)		<−70		
Dielectric constant at 1 MHz		2.35		

Source: BP Chemicals (UK) Ltd. (quoted in Wolpert, *Synthetic and Plastics Paper*, Vol. 1, 1971, p. 17).

In the case of medium-molecular-weight HD PE, Hoechst has been offering Hostalen grade of GF 7750 M, which has the properties shown in Table 5.16. Paper-like film can be extruded from granules or powder.

Hoechst has continued its R&D work on producing a grade with better melt-flow characteristics. In autumn of 1975 they developed the new grade of high-molecular-weight HD PE—Hostalen GMVP 9255F—with a wider range of molecular-weight distribution, a specific density of 0.950–0.954 g/cm^3, a melting point range of 125–133°C, and a melt index (MFI 190/5) of 0.2–0.4 g/10 min. This grade represents an improvement over the previously supplied grades, an improvement which has been confirmed to the author of this book by several extruder manufacturers and plastics paper makers that have been using the new grade for trial runs.

Dow Chemical Grade XP 3574.12

Dow Chemical has been building up its HD PE production capacity in Europe and has developed grade XP 3574.12 of 0.950-g/cm^3 density with

Table 5.15 *Properties of High-Molecular-Weight Hostalen G*

Property	*GM 9955 F2*	*LP 103*
Melt index	0.51-0.65	0.5-0.75
Specific weight	0.952-0.957	0.951-0.956

Source: Hoechst AG, Fed Rep Germany (quoted in Wolpert, *Synthetic and Plastics Paper*, Vol. 4, 1974, p. 620).

Table 5.16 *Properties of Medium-Molecular-Weight Hostalen Grade GF 7750M*

Property	*Unit*	*Value*	*Testing method*
Specific weight	g/cm^3	0.948	DIN 53 479
Melt index MFI 190/5	g/10 min	1.9	DIN 53 735 E

Source: Hoechst AG, Fed Rep Germany (quoted in Wolpert, *Synthetic and Plastics Paper*, Vol. 4, 1974, p. 620).

a melt index of 0.25 (190/2.16) for making plastics paper. The company offers this polymer in microgranules, since it feels that extrusion from microgranules offers additional advantages.

Lupolen

BASF (Fed Rep Germany) has been offering some grades of their Lupolen HD PE polymers for making paper-like film. Grade 4261 AX of 0.945 density is supplied in the form of uncolored "grit." After annealing for one hour at 100°C, the density of this high-molecular-weight grade is 0.943–0.946, whereas grade 5261, also supplied in "grit" form, has a slightly higher density of 0.950–0.953.

Vestolen A

Vestolen A, manufactured by Vestolen GmbH (Fed Rep Germany), a joint venture of Chemishce Werke Hüls and VEBA-Chemie, and marketed by Hüls, is made by the Ziegler process. This HD PE polymer for making paper-like HD PE film is supplied in several grades, including grade 6042P which has the properties shown in Table 5.17.

Grade AX 2041 has a higher molecular weight than 6042P and has been developed for a film to make bags of greater strength. It has a density of 0.955 and a melt-flow index (MI 5) of 0.5–0.7.

Wacker-Chemie Grades of HD PE

Wacker-Chemie of Munich (Fed Rep Germany) started manufacturing HD PE polymers in 1971 and has been concentrating on the development of several grades for making paper-like films since the company is confident that this market will be growing at a high rate.

Among its grades of high-molecular-weight HD PE are DF 5071G, DF 5074G, and DF 5078G, all with a specific density of 0.952. Wacker-

Table 5.17 *Properties of Vestolen A Grade 6042P*

Property	*Value*
Specific density	0.958
Melt flow index (MI 2.16)	0.2-0.3
Melt flow index (MI 5)	1.0-1.4
Melting range (°C)	132-135
Softening point (°C)	122
Vicat softening point (°C)	67

Source: Vestolen GmbH, Fed Rep Germany (quoted in Wolpert, *Synthetic and Plastics Paper*, Vol. 1, 1971, p. 23).

Chemie has been supplying some grades with antistatic properties. For film for making carrier bags the company supplies polymers containing 5% titanium dioxide. In addition, it has developed grades of medium-molecular-weight HD PE (e.g., FF 5551G of 0.955 specific density) for the applications where the use of a lower-molecular-weight HD PE is more economical. At the K-75 exhibition at Dusseldorf (October 1975) Wacker exhibited samples of printed paper extruded from a compound of HD PE filled with calcium carbonate.

Sclair

This HD PE polymer is manufactured by Du Pont of Canada Ltd., which developed grade 19A for making paper-like film in 1968,[17] and later added Sclair 17A.[18] This grade was specially developed for making film for paper and tissue replacement and for laminates as well as for coextruded applications. To assure consistent high-quality film production, Sclair 17A is continuously monitored for gel.

According to Du Pont, because of the sensitivity of Sclair 17A to orientation effects, the property balance attained is dependent on the blow-up and draw-down ratios. For making HD PE film applications, the optimum-blow-up ratio has proved to be from 3:1 to 5:1.[19] The physical properties given in Table 5.18 were obtained at a melt temperature of 400°F (205°C) and are typical of Sclair 17A blown into film on an adjustable-tower system.

17. *Synthetic and Plastics Paper*, Vol. 1, pp. 33-37.

18. Ibid, Vol. 2, pp. 203-4.

19. Since the first introduction of paper-like HD PE film, there has been an overall trend towards a reduced blow-up ratio.

Table 5.18 *Properties of Sclair 17A*

Property	*ASTM test method*	*Values*	
		Metric units	*British units*
Melt index	D1238-65T	0.40 dg/min	0.40
Film density	D1505-68	0.950 g/cc	0.950
Film gauge		30 microns	1.1 mils
Blow-up ratio		4:1	4:1
Draw-down ratio		4:1	4:1
Tensile properties, MD	D638-68		
Tensile yield		265.0 kg/cm^2	3,750 psi
Ultimate tensile strength		320.0 kg/cm^2	4,530 psi
Elongation		550%	550%
Tensile properties, TD	D638-68		
Tensile yield		270.0 kg/cm^2	3,830 psi
Ultimate tensile strength		250.0 kg/cm^2	3,530 psi
Elongation		275%	275%
Elmendorf tear strength	D1922		
MD		1.0 g/micron	23 g/mil
TD		9.0 g/micron	220 g/mil
Tensile modulus	D638-68		
MD		7,920.0 kg/cm^2	113,000 psi
TD		10,530.0 kg/cm^2	150,000 psi
Vicat softening point	D1525-65T	123°C	253°F

Source: Wolpert, *Synthetic and Plastics Paper*, Vol. 2, 1972, p. 204, based on information supplied by Du Pont of Canada, Montreal.

Chemplex

This HD PE homopolymer, manufactured by Chemplex Company (USA), is supplied in several grades: grade 6108 provides the stiffest paper-like film; grade 5704, the toughest film, and grade 5103, paper-like film with properties intermediate between the first two grades. Tables 5.19 and 5.20 show the physical properties of these three grades and the properties of film blown from Chemplex 6108.

Fortiflex

Fortiflex, produced by Celanese Corporation of America (USA) under Phillips license, has been marketed by Amcel Co. Inc. (an affiliated company) in the USA and in Scandinavian and other countries. The various grades are supplied in a range of densities (0.957–0.960) for making tissuelike plastics paper. At trials with its grade F-141A on Rotatruder equipment (Oerlikon-Bührle), film of slightly over 2-micron

Table 5.19 *Chemplex Polyethylene – Typical Physical Properties of HD PE Film Resins*

Property	*Unit*	*ASTM test method*	*Grade* 6108	5103	5704
Melt index	g/10 min	D 1238	0.8	0.3	0.8
Density	g/cc	D 1505	0.960	0.950	0.954
Tensile yield	psi	D 638*	4,000	3,300	4,400
Flexural modulus	psi	D 790	220,000	165,000	150,000
Impact strength	ft · lb/in	D 256	4	2	6
Brittleness temperature	°F	D 746	<180	<180	<180

* Using Bar 'C' of ASTM D 412 at 2"/min testing speed.

Source: Wolpert, *Synthetic and Plastics Paper*, Vol. 2, 1972, p. 201, based on information supplied by Chemplex.

Table 5.20 *Typical Properties of HD PE Film*

Property		*Film thickness (mil)* 0.6	1.0	2.0	3.0
Elmendorf tear (g/mil)	MD	<1	6	18	75
	TD	707	87	34	36
TMI ball impact (cm-kg)		0.6	0.9	2.3	3.8
Tensile properties (ASTM D 882, 20 in/min)					
Yield (psi)	MD	4,780	4,650	4,620	4,690
	TD	4,400	4,170	4,230	4,290
Break (psi)	MD	5,690	5,090	3,630	3,520
	TD	4,040	4,000	3,420	2,980
Elongation (%)	MD	430	565	155	54
	TD	42	26	165	105
Modulus, 1% secant (psi)	MD	184,000	172,000	169,000	178,000
	TD	214,000	198,000	188,000	187,000

Note: Film blown from Chemplex Polyethylene 6108 – blow-up ratio 3.3:1. From thinnest to thickest film, draw ratios were 15:1, 9:1, 4.5:1 and 3:1.

Source: Wolpert, *Synthetic and Plastics Paper*, Vol. 2, 1972, p. 202.

Table 5.21 *Properties of Moplen-RO*

Property	*ASTM test method*	*ZB5015*	*ZB6015*
Nominal density		0.950	0.960
Melt index (g/10 min)	D1238-65T, E	0.3	0.3
Flexural stiffness (kg/cm^2)	D747-63	8,500	12,000
Vicat softening temp (°C) (1 kg in air)	D152-65T	132	135

Source: Wolpert, *Synthetic and Plastics Paper*, Vol. 1, 1971, p. 26.

thickness was extruded. The HD PE manufacturing plant of Celanese Corporation was recently purchased by Solvay.

Moplen-RO

Moplen-RO, manufactured by Montedison (Italy), is supplied for making tissuelike plastics paper. Montedison's HD PE grades have the properties given in Table 5.21. The characteristics of grade ZB 5015 are good flexibility, high impact strength, and high durability, whereas the main characteristic of the grade ZB 6015 is rigidity. These grades can be used together as compounds.

Rumiten

This HD PE polymer of 0.960 density is manufactured by Rumianca SpA (Italy). Type 615 is a homopolymer with superior impact strength; type 630 is a medium-flow homopolymer; and type 650 is described as a high-flow homopolymer.

Manolene

This HD PE polymer has several special grades for making plastics paper. It is manufactured in France and Belgium by joint enterprises of Phillips and Rhone-Poulenc, using the Phillips process, and is marketed by Sodefine, Paris.

Natene

Natene, developed and manufactured by Pechiney-Saint-Gobain, France, has the properties given in Table 5.22.

Table 5.22 *Properties of Natene*

Property	*Standard*	*Unit*	*45021 AA*	*60061 AA **
Melt index	ASTM D 1238/65T	g/10 min	0.2	0.6
Density	ASTM D 792/66T		0.945	0.960
Stiffness	ISO R 527	kg/cm^2	6,000	12,000
Flexural strength	NFT 51-001	kg/cm^2	450	640
Elongation at break	ISO R 527	%	1,100	1,100
Vicat softening temp (under 1-kg load)		°C	115	126
Melting temp		°C	124	133

Source: Wolpert, *Synthetic and Plastics Paper*, Vol. 1, 1971, p. 25.

* This grade is for films with a greater stiffness.

Staflene

This HD PE polymer is produced by applying the medium-pressure process of Standard Oil Company (Indiana) and is manufactured by Furukawa Chemical Industries (Japan). For blown-film extrusion, Staflene is supplied in a range of grades with a melt index of 0.3–2.0. Grade E 615 is a homopolymer of 0.96 density and of 1.5 melt index. Grade E 505 is an ethylene-propylene-copolymer of 0.95 density and of 0.5 melt index; it is mainly used for making microthin film.

New Process and Catalysts for Making HD PE Polymers

The Gas-phase Process

Union Carbide Corporation has developed the gas-phase process for making HD PE. This process uses special catalysts and eliminates the solvent recovery stage or discharge of effluent. According to UCC, the process reduces the production costs of HD PE because of lower equipment and operating costs.

Although some manufacturers feel that the range of HD PE polymers made by this process is limited, it is significant that the Unifos (Unifos Kemi AB) plant in Sweden, using this process, has developed several grades of HD PE of 0.962 density with different melt index properties for making films replacing tissuetype paper, glassine, and greaseproof paper. In addition, the company's grade DGDJ-6090 of 0.960 density, with a melt index of 0.3, has been developed for making a tough film to replace kraft paper. With the growing demand of plastics paper, it is feasible that 15,000-tpa reactors will be used exclusively for production of those grades of HD PE, from which films will be made to replace conventional paper.

In addition to the Swedish plant, which has exceeded the designed 32,000-tpa rate, this gas-phase process is applied in some other plants in operation or under construction in the USA, the USSR, Australia, Czechoslovakia, and other countries.

The patent "Process for Polymerization" granted to Union Carbide refers to supported diarene chromium compounds as the catalysts for catalytic polymerization of olefins.[20] The UCC invention says that when these diarene compounds are modified with trace amounts of cyclopentadiene, it improves their behavior in response to hydrogen as a melt-index control agent, and the molecular nature of the olefin polymers obtained is also improved.[21]

New Catalysts

Mitsubishi of Japan has incorporated a new catalyst in the company's 60,000-tpa HD PE plant, which has been on-stream since 1972. It is claimed that this catalyst lowers the production costs of the polymer and that interest in this process has been shown in various countries. Other polymer manufacturers, including Hoechst, Montedison, and Solvay, have also succeeded in developing new catalysts.

EXTRUDERS FOR MAKING PLASTICS PAPER

Leading manufacturers of extrusion equipment for making LD PE film, realizing the great potential of the market, have developed modified equipment to extrude plastics paper.

Plastics paper is an extruded HD PE film, and leading HD PE resin manufacturers have developed special grades of this polymer for this application. The extruded film does not basically require any additional treatment. There are two basic methods of extrusion:

1. Cast-film extrusion. The polymer melt is extruded through a flat film die and then chilled.
2. Blown-film extrusion. The polymer melt is forced through an annular (circular) die and a tube is formed which is inflated (blown) to the desired diameter. In this way the final thickness of the film is controlled. The inflated tubular film passes through a cooling tower and is

20. British Patent 1,317,734, Complete Specification published May 23, 1973.

21. Ibid.

then flattened and wound up. In the blown-film process the comparatively thick walled extruded tube is biaxially oriented to a thin film. The longitudinal orientation is controlled by the ratio or the extrusion rate and the speed of the pinch-rolls. The transverse orientation is controlled by the ratio between the final tube diameter and the initial tube diameter.

At present, plastics paper is produced exclusively by the blown-film extrusion technique, but it is possible that cast-film (flat die) methods will be applied some time in the future, particularly when lower tolerance variations are a requirement (for example, in electrical-industry applications).

Manufacturers' Comments

Bone Cravens

K.J. Braun, of Bone Cravens Ltd. (UK), manufacturers of extrusion equipment, in his presentation at the "Plastics and Paper" conference in London (May 1973) said that

> The main external difference between plants for low-density polyethylene and high-density polyethylene film is the height of the nip rolls. For high-density polyethylene, a typical distance from the die to the nip rolls is 2–3 m. This distance is not a fundamental parameter, but the relatively low height allows the film to collapse whilst still warm and pliable, thus reducing the tendency to crease. Since HD PE film has little tendency to block, this process has the advantage of requiring less height than corresponding plant for LD PE.
>
> The other feature of HD PE film plant which differs from the corresponding LD PE film plant is the blow-up ratio (ratio of bubble diameter to the die diameter) normally used, and this for bag applications tends to be in the range between 3:1 and 6:1, but may be as low as 2:1 for 'tissue' applications. The choice of the blow-up ratio depends on the required properties of the film; the transverse orientation increases with the blow-up ratio, but this ratio is by no means the only important parameter governing the properties of HD PE film. The polymer itself has the greatest bearing on this, but bubble shape can often produce a more marked effect on the mechanical properties of the film than the blow-up ratio.
>
> The mechanical design of the extruder and ancillary components must take into account the very high processing viscosities of

some of the high-molecular weight grades of HD PE, and consequently the torque-capacity of the drive and the rating of the main extruder thrust bearing must be capable of withstanding the associated heavy loads, as must all the other components of the extrusion system which are subjected to the effects of the high pressure . . . .

The die must be mechanically designed to withstand the high pressure generated within it, whilst at the same time the pressure drop through the die should be maintained at a minimum, consistent with good quality of film, as any excessive pressure results in a higher heat content of the melt leaving the die, and thus limits the output of the extruder.

In order to obliterate the memory effects caused by the elastic content of the melt, the die must be designed to produce a sufficient amount of mixing and relaxation of the melt and as uniform a shear history as possible. For this reason, the trend in the die design has been away from the side fed across head type or a simple form of spider die.

Braun also said that a spiral mandrel die has been applied successfully. The output rate depends to a large extent on the polymer, and Braun referred to the production rates in the case of different die sizes and different HD PE types, using the Bone Cravens 60-mm Powerflight extruder (Table 5.23). He added that with higher web widths and larger extruders (90 mm) production rates in excess of 100 kg/hr were achieved.

Paul Troester

Rolf Kirchhof, of Paul Troester Maschinenfabrik (Fed Rep Germany), referred in his presentation at the IUPAC-EUCEPA Conference in Helsinki (1972) to a U.S. paper manufacturer who has carried out trials with a 4½-in extruder with a 500-mm ring-gap diameter (film of 2,000-mm tube circumference), achieving an output of 200 kg/hr of tissue-type plastics paper. Kirchhof felt that extruders of larger sizes would considerably lower the production cost of plastics paper and referred to the growing interest in 6-in (150 mm) extruders. He observed that it was possible that flat-die extrusion equipment will be used in the future for making HD PE films. During the discussion Kirchhof said the company had observed thickness tolerances on its blow-film extruders of plus/minus 5% in a film of about 25 microns. These tolerances referred to across-the-film measurement, since the longitudinal tolerances were only plus/minus 2% and were dependent on the screw rotating speed.

Table 5.23 *Performance Summary of 1-Meter HD PE Film Plant – 2½-in "Powerflight" Extruder*

Polymer	*Layflat width (mm)*	*Die size (mm)*	*Output (kg/hr)*	*Thickness (micron)*
(1) Hostalen 9955 F*	1,000	150	80	15-20
	500	75	50	15-20
(2) Shell 040†	1,000	150	65	15-20
	1,000	125	55	15-20
	500	75	50	15-20
(2) Shell EB183†	1,000	200	100	10
(3) Rigidex Type 9‡	1,000	200	95	10

Source: K. J. Braun, Bone Cravens Ltd., "Extruded HD PE Paper Production: Technology and Economics" (Paper presented at the Plastics and Paper Conference, London, May 1973).

Note: "Powerflight" extruders are built by Bone Cravens Ltd. (UK).

* Manufactured by Farbwerke Hoechst, Fed Rep Germany.

† Supplied by Shell Chemicals, UK.

‡ Manufactured by BP Chemicals, UK.

Oerlikon-Bührle

According to A. Petris,[22] Oerlikon-Bührle, HD PE production lines with vertical extruders, oscillating 360 degrees, have proved particularly successful in practice; the main advantages of this extrusion system are:

1. Symmetrical and constant pressure flow of material without deflection from barrel into die. Thus gauge tolerances, caused by the die, can be kept in a close compass and are independent of screw speed.
2. No sealing of rotating parts in the melting and high-pressure zones is required. This results in space saving and compact design, and makes it possible for the negative effects on gauge tolerances of the film on the reel to be eliminated by impeding a deposit of the melt.

Petris referred to the following production figures for HD PE film of 25-micron thickness, which indicate recent progress achieved with standard plant incorporating a 60-mm screw extruder.

Die	*Production Capacity*
115-mm dia	70 kg/h
200-mm dia	100 kg/h

22. A. Petris, "Paper-like Film," (Presentation at TAPPI Synthetic Papers & Pulps International Symposium, Washington, D.C., March 31 - April 2, 1976).

Extruder Development

Bone Cravens

Bone Cravens' latest extrusion plant includes a rotating haul-off unit and winder in order to achieve better reels of film, which is important in the further converting of this material. Bone Cravens' plant for making plastics paper (HD PE tubular film of a maximum width of 1000 mm), shown at K-75 exhibition at Dusseldorf (October 1975), incorporated the following new features:

- □ Semi-automatic reel change, fitted to each of the two winding stations, resulting in lower production-plant labor requirements.
- □ New width control system, which automatically inflates and deflates the film bubble, in order to maintain a constant layflat width of the film (tolerance: plus/minus 1 mm). The unit is also fitted with a digital set width for start-up operation, and enables a rapid change of width when required.

This plant, with an overall height of 6,850 mm, incorporates a 60-mm extruder of 24:1 L/D ratio. The extruder barrel is fitted with heaters/coolers in four separate zones, and the screw is of a new design with a 2-stage mixing section in order to achieve a high output of homogeneous melt at a low melt temperature, and a high film-production rate. The 150-mm extrusion die is of spiral mandrel design, and the labyrinth-design cooling ring is mounted immediately above the extrusion die to cool the film bubble, since it extrudes vertically upward from the die. The plant also incorporates a rotating haul-off, a slitting station, a film guidance unit and a gusseting unit. On both sides of the film take-off unit is a surface winder. Among optional extras, Bone Cravens offers a high-voltage Corona discharge unit to treat the film on both sides and to provide a better printing-ink adhesion.

Oerlikon-Bührle

Oerlikon-Bührle of Zurich (Switzerland), manufacturer of the well-known Rotatruder extruders in which the cylinder, screw, head, and cooling ring are mounted on a 360° rotating platform, has developed a special mixing-zone screw in order to achieve a symmetrical melt flow. The company has also developed a thermal homogenization of the melt in order to prevent the otherwise-occurring difference of over 25°C in the inner/outer temperature of the melt at the extrusion point.

Bone Cravens 1-m rotating plastics paper plant based on a 60-mm Powerflight extruder and fitted with two winders. Courtesy Bone Cravens.

Bone Cravens plastics paper plant, showing the HD PE film bubble and the cooling ring guides. Courtesy Bone Cravens.

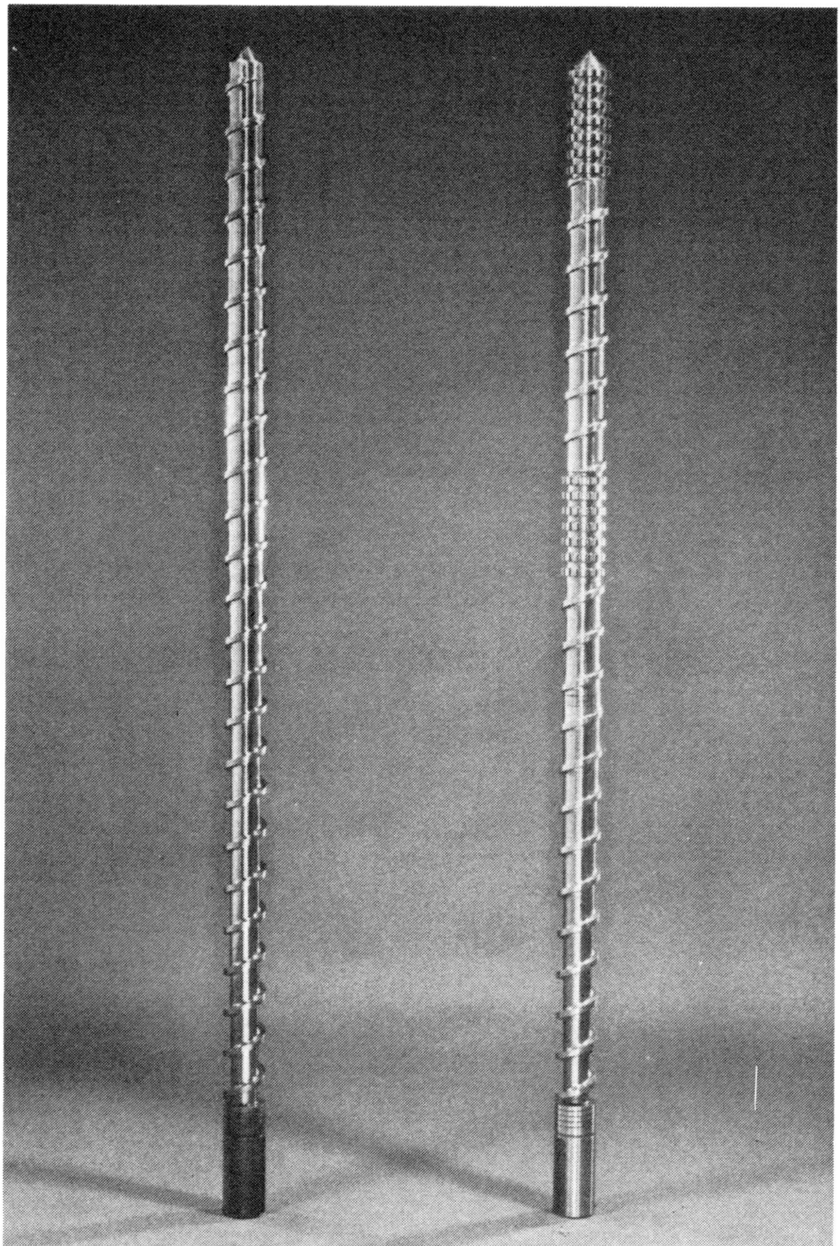

Oerlikon-Bührle's special mixing-zone screw is shown at right. Courtesy Oerlikon-Bührle.

Oerlikon-Bührle plastics paper production plant with 60-mm Rotatruder (VO6).

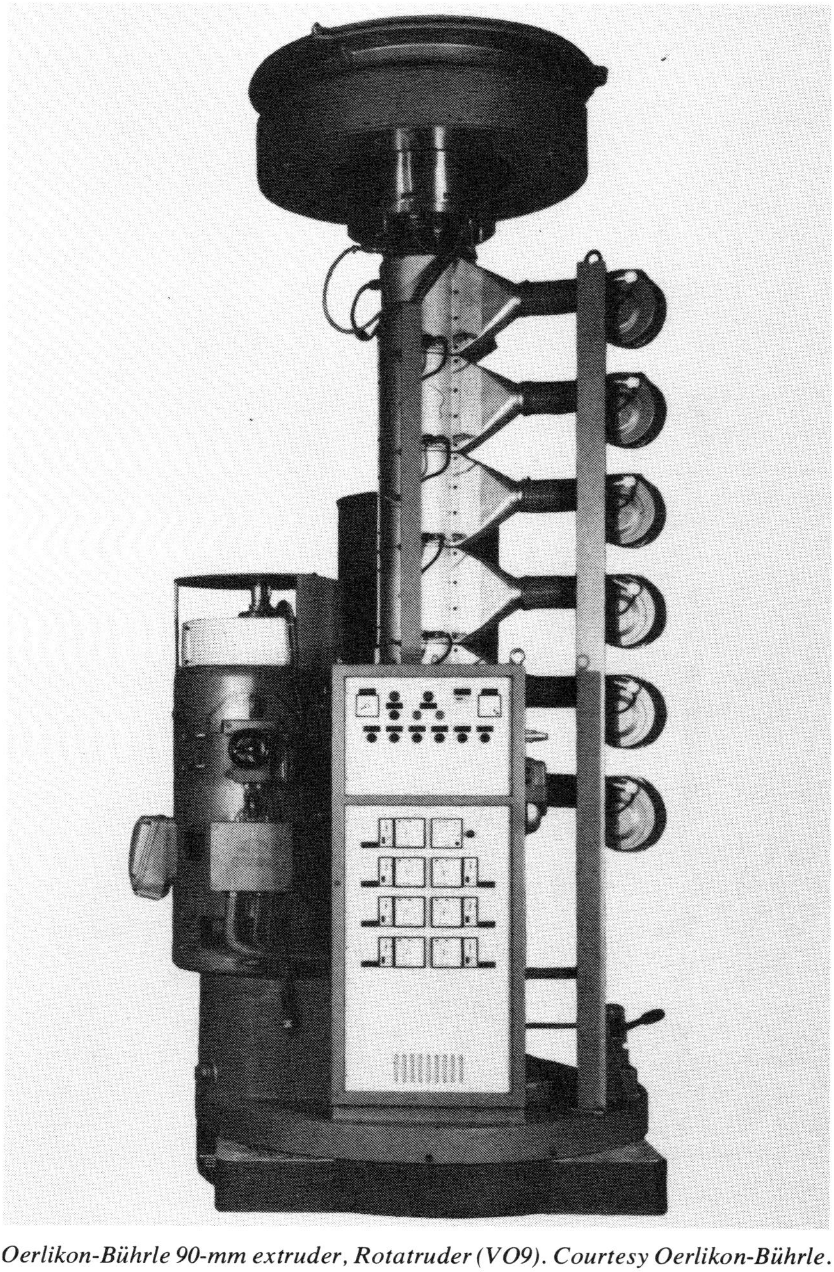

Oerlikon-Bührle 90-mm extruder, Rotatruder (VO9). Courtesy Oerlikon-Bührle.

Table 5.24 *Production Rates for Rotatruder RO 60 (November 1974)*

Die	*High-molecular-weight HD PE*	*Medium-molecular-weight HD PE*
115 mm	60 kg/hr	80 kg/hr
130 mm	70 kg/hr	90 kg/hr
150 mm	80 kg/hr	100 kg/hr

Notes: All values are approximate.

Output also depends on the raw material and dimension.

Oerlikon feels that the combined take-off winder equipment is of great importance in the production of plastics paper. The company supplies a contact winder, type WKB, which is driven by a variable-speed transmission that drives the axis of the nip roll directly. The same transmission drives the contact roll and the guide roll over sliding clutches. The tension of the film between the take-off unit and the contact winder is regulated by the nip-roll drive. The cut tension, which is necessary for the cutting device, can be regulated independently. An antiwrinkle expander is installed for the production of tissue-type plastics paper. Among the most recent Oerlikon developments is the introduction of the floor-level winding unit, that is, a unit in which both the extruder and the winding unit (featuring fast reel change) are on the same level.[23]

At a demonstration (November 1974) of extruding 20-micron plastics paper on Rotatruder RO 60 (i.e., 60 mm) from high-molecular-weight HD PE produced by Wacker Chemie München, manufacturers of several types of high-molecular- and lower-molecular-weight HD PE, Oerlikon quoted the production rates for its 60-mm extruder. These are shown in Table 5.24. In the case of a 90-mm extruder correspondingly higher output rates are achieved.

Reifenhäuser

In 1973 Reifenhäuser KG (Fed Rep Germany), a leading supplier of extrusion equipment, demonstrated four standard tubular extrusion lines, including a line with a 90-mm screw diameter for making paper-like HD PE film. On all four lines the film had a thickness tolerance of plus/minus 15%.

23. *Synthetic and Plastics Paper*, Vol. 4, p. 631.

Line 1: 30-mm screw diameter, throughput = 20 to 22 kg/hr; for making folded bags from 17- to 19-micron film, in sizes between 105-mm lay-out width, 25-mm gusset depth, to 200-mm lay-out width and 40-mm gusset depth; flat small shopping bags with lay-out width from 150 mm to 190 mm, and length of 250 mm to 270 mm respectively. Grade GMVP 9955 F of Hoechst was extruded.

Line 2: 60-mm screw diameter, rotating die-head system, tube circumference = 144-mm, throughput = 70 kg/hr (die 80-mm ring-gap diameter), film thickness 20 microns, extruder and winder on the same level. During demonstration grade GMVP 9955 F of Hoechst was extruded.

Line 3: 60-mm screw diameter extruder, rotating haul-off and winding combination system, tube circumference = 1520 mm, throughput = 80 kg/hr (die 150-mm ring-gap diameter), 10-micron film, extruder and winder on separate levels. Grade GF 4760 of Hoechst was extruded.

Line 4: Extruder type ET 650-1-90-25 with a 90-mm screw diameter, tube circumference = 2400 mm, film thickness: 20 micron, throughput = 110 kg/hr (die 150-mm ring-gap diameter), extruder and winder on separate levels. Grade GMVP 9955 F of Hoechst was extruded.

Ing. H. Rahlfs and Ing. W. Dreser of Reifenhaüser, in their joint presentation at this demonstration, stressed the point that "High molecular weight materials having a low melt index require plastification which does not overstrain the material . . . a so-called decompression screw has considerable advantages over conventional 3-zone screws, as it gives a lower melt temperature over the whole processing range with even better homogeneity of the melt."

They compared two basic design systems for extrusion plant:

System "A", in which the extruder die head and air ring are stationary (i.e. non-rotating), but the bubble collapsing guides which are below a combined take-off and winder system are adjustable in height and rotate as an entity.

System "B", in which extruder, bubble collapsing guides, height adjustable take-off and winders are all non-rotating. The die head and the air ring are rotating. The winder is at the same level as the extruder, and thus only one working level is required.

They felt that the advantages of system "A" are:

a) parallel cylindrical reels,
b) satisfactory printing properties,
c) satisfactory properties for laminating,

and due to the above advantages, system "A" offers far greater reliability with regard to reel quality and behaviour during running of the plant in production.

System "B" has the following advantages:

a) only one working level,
b) lower overall height,
c) longer cooling distance for the film until reaching the winders.

System "B" has the following disadvantages:

The system produces relatively good looking reels, but uniformity of reel circumference, i.e. the reel being parallel, is not definitely assured, and it will be found that deviations in reel diameter exist from reel to reel.

In system "B", any errors which might arise from the rotating die head and air ring are camouflaged, but on the other hand errors from external sources, such as ambient conditions, will always appear in the same position on the film reel.[24]

Reifenhäuser has redesigned its extrusion lines for making plastics paper, and at the end of 1975 started offering 50-mm and 70-mm diameter screw extruders, with the screw of a special design, having a feedzone with grooves and an intense cooling arrangement. The extruders are supplied with barrels of different lengths: 15D, 18D, 20D, and 25D. Die inserts with diameters of 80, 100, 120, and 150 mm are interchangeable on these extruders.

Reifenhäuser recommends that the extruders should be operated vertically downward, with the reelstand on floor level, and stresses the advantage of using the rotating take-off system (rotating bubble). The take-off equipment is supplied in three sizes (900 mm, 1,300 mm and 1,650 mm), and the largest size is of particular interest to manufacturers that can convert plastics paper by slit-sealers and center-folding units. In addition to these extruders of new design, the company will continue supplying the conventional plant equipped with 60-mm and 90-mm extruders together with the haul-off and winder combination, mounted on a tower, the platform of which is adjustable in height.

Extruder Manufacturers at the K-75 Plastics Industry Exhibition

It is significant that at the K-75 plastics industry exhibition in Dusseldorf (October 1975) more than 10 manufacturers were offering machin-

24. H. Rahlfs and W. Dreser joint presentation at the Reifenhäuser extrusion plant demonstration, Troisdorf, February 1973, quoted in *Synthetic and Plastics Paper*, Vol. 4, pp. 628-29.

Reifenhäuser 50-mm extrusion plant for making plastics paper. Courtesy Reifenhäuser.

ery for making plastics paper. Furthermore, improved equipment for making plain and printed bags from plastics paper was on show. Kiefel (Fed Rep Germany) adiabatic extruders have an L/D ratio of 12:1 and operate without a heat supply, thus having considerably low energy consumption and a comparatively low price. These Kiefel extruders have mainly been installed for making thin plastics paper for wrapping applications. Most extruder manufacturers, however, supply polytropic extruders with a higher L/D ratio, since these extruders allow better control and influence of the temperature and can, therefore, be better adjusted to the various types of polymer and individual charges. Alpine (Fed Rep Germany) makes extruders with an L/D ratio of 16:1 and screw diameters of 35 and 50 mm. The company offers extruders with either upward or downward extrusion. Bielloni (Italy) offers 'HD' extrusion equipment with the very high L/D ratio of 33:1, and Dolci (Italy) reports having secured orders for HD PE extrusion equipment. Crespi Costruzioni Meccaniche (Italy) offers twin-head extruders. Machinery exhibitors pointed out they were receiving orders from plastics paper manufacturers who were expanding their production capacity, and from European and other newcomers in this field.

Economics of Production Cost

During the introduction period, while the market was tested, it was felt that 60-mm extruders, and even some smaller extruders, were the appropriate ones to use because of their comparatively low investment cost.

The first product to be introduced was tissue-type plastics paper, mainly for hand-wrapping of meat, wet fish, flowers, sandwiches, vegetables, etc. This plastics paper of about 10-micron thickness (later of even lower gauges) has the required properties of being greaseproof, water-proof, and oil-resistant and has a low gas permeability, *but* it does not require high tear strength. It is therefore more economical to use lower-molecular-weight polymers in its production than high-molecular-weight HD PE, and still produce a material with the required properties. Using lower-molecular-weight HD PE reduces cost in two ways:

1. The polymer is cheaper.
2. The output rate of film from lower-molecular-weight HD PE is about 25% to 30% higher than that from high-molecular-weight HD PE, using the same size extruder.

Some converters producing tissue-type paper-like HD PE film have installed adiabatic extruders made by Paul Kiefel of Worms (Fed Rep Germany). These extruders are considerably cheaper than conventional extruders. The adiabatic extruders operate without heat supply and heat discharge and have a V-belt drive. The screw has an L/D ratio of 12:1, as against the usual L/D ratio of 24:1, and has been supplied in various sizes, starting with a 17-mm diameter.

Following the introduction of tissue-type plastics paper a much bigger market (in terms of tons) has developed for plastics paper for bags. In addition to all the requirements of tissue-type plastics paper, the material for bags requires a higher tensile strength. For retail bags, plastics paper of about 20- to 25-micron thickness is being used, while for other bags (e.g., gusseted free-standing shopping bags) plastics paper of higher thickness, in some cases 70 microns, is required. For bags which require a higher tensile strength, it is necessary to use film made from high-molecular-weight HD PE.

There are two schools of thought on economics in respect to extruders for making this type of plastics paper:

1. The installation of several extrusion lines with 60-mm (or even smaller) extruders, each line producing a specific gauge, and followed up with a bag-making unit producing a specific size of bags, provides a greater flexibility than the installation of high-output 90-mm or 120-mm extruders, which may be using the multilane technique. The switch-over from one gauge to another results in the loss of working time of expensive equipment.
2. The large high-output extruders provide a considerable saving in production cost, including labor cost, despite the initial higher investment that is required. Large-scale converters, including paper manufacturers who have been diversifying their activities by entering the field of plastics paper production, feel that this type of equipment enables them to compete successfully with smaller converters who cannot afford this type of investment.

It is estimated that the production costs account for about 30-40% and the cost of polymers for about 60-70% of the total cost of plastics paper.

The proper choice of the size of the equipment depends on the size of the present and future market that the plastics paper manufacturer expects to secure, and on the range of different types of plastics paper, and their thicknesses, he intends to produce. A further consideration is whether he could utilize this equipment, if necessary, for production of other types of films, particularly at the initial stage.

Gloenco Ltd. of Britain, which is a subsidiary of Gloucester Engineering (USA) and which manufactures extrusion equipment, has developed 45-mm extruder lines for making bags at a rate of only 40 kg/hr, since they felt that at the time when the market was comparatively small it was more economical to install three or four such lines, each running specific sizes of bags, coupled with a printing press. Lately, however, they have been active in supplying complete production lines incorporating 90-mm extruders.

Developing countries that have to import paper bags or paper for making bags have been showing increasing interest in installing production lines for making plastics paper and bags from this paper at comparatively low investment costs.

6. Oil Prices Impact and Overall Considerations

PRICE DEVELOPMENTS SINCE SEPTEMBER 1973

The increased prices of crude oil since the end of 1973 have had strong repercussions on world economy, international trade, and foreign-trade balances of individual countries. The chapter of history characterized by an abundant supply of cheap crude oil, on which the world economy was to some extent based, has ended. This means that industrialized and developing countries will have to find ways to deal with the newly created conditions.

It is beyond the scope of this book to discuss the complex oil situation within the total concept of energy supply and oil requirements, or to compare, within the overall inflation picture, the changes in price of various commodities and industrial products with those of oil during the decade preceding the end of 1973. This chapter deals instead with the main implications of increased oil prices on the synthetic polymer and the traditional pulp/paper industries, and tries to evaluate the competitive strength of synthetic polymers on the basis of these increases.

At present the synthetic polymer industry uses oil as a feedstock and as an energy source. The consumption of the synthetic polymer industry (synthetic fibers, plastics, and other synthetic polymer products) accounts for about 2% of oil consumption worldwide. Since the products of this industry are highly up-valued crude-oil products, it can be assumed that the industry will be able to secure the increasing required amounts of this raw material, particularly since there is a growing recognition that the burning of oil is a wasteful use of this material. This up-valuing of crude oil by processing it to naphtha and monomers, and possibly even to polymers, is one of the main reasons for the interest of oil-exporting countries in establishing indigenous petrochemical and polymer industries.

Table 6.1 *Production Cost of Ethylene*

Thousand tons/yr capacity	100	200	300	400	500
Total cost £/ton ethylene	58	48	42	39	37

Note: All at same unit feedstock cost and same unit prices for products and at full load.

Source: Shell Oil, "Petroleum-based Chemicals – the Feedstock Problem" (London, October, 1973).

In its publication "Petroleum-based chemicals—the feedstock problem" (issued in October 1973) Shell Oil points out that the technology of the oil-based chemical industry was developed to use the low-cost feedstocks, and that in using them the industry gave them a very much higher value than they could command as sources of energy. Shell also points out that since World War II petroleum-based chemicals have been among the fastest growing sectors of the U.S., European, and Japanese economies. The publication shows that the unit cost of ethylene production drops with the increased size of ethylene plants, and it thus indicates the dilution of the raw material factor in the total production cost of ethylene, as shown in Table 6.1.

In Europe, particularly, the growing demand for naphtha as ethylene feedstock or as feed to aromatics plants is now in direct competition with its alternative use as a gasoline component. Table 6.2 compares demand in Europe and the USA. The U.S. gasoline market takes a greater share of total consumption of crude oil and is large compared with the size and growth of the gas-oil (or heating) market.

Another difference between the USA and Europe is that oil refining in the USA is based to a greater extent on thermal and catalytic cracking than it is in Europe. This results in a higher proportion of propylene and other light olefins, which can be used as chemical feedstocks. In fact, out of 100 tons of propylene used in chemicals production, 70 are from

Table 6.2 *Product Demand Pattern (as % of Ton of Crude Oil)*

	USA	*Europe*
Gasoline	46	25
Middle distillates	25	32
Rest	29	43

Source: Shell Oil, "Petroleum-based Chemicals – the Feedstock Problem" (London, October 1973).

catalytic crackers and 30 from ethylene plants in the USA, whereas in Europe the ratio is almost exactly the reverse.

According to Shell the U.S. chemical industry is now following the European example of producing olefins from naphtha and middle distillates, and this is part of a worldwide trend to utilize heavier feedstocks for oil-based chemicals production. If European chemical companies are to continue this trend towards the cracking of higher oil fractions, they will have to crack very heavy, high-sulphur feedstocks, because these are the only parts of the oil barrel in Europe that are now likely to be freely available at a relatively low price. In this way, the industry would be able to find advantages in optimizing combinations of refinery processes and base-chemical plants.

Shell stresses the importance of improving the utilization of some chemical feedstocks, and of new technologies enabling the use of these products of oil operations which cost relatively little and are freely available:

> There are sufficient hydrocarbon feedstocks physically available to meet all possible chemical demands. Even by 1980 chemical demand is expected to rise only to 5.5% of the world's total demand for hydrocarbons. But in order to be allowed to use these feedstocks the chemicals industry must demonstrate that the chemical use of hydrocarbons will bring better returns than the application of these hydrocarbons as energy sources.[1]

In this context reference should be made to the production of ethylene by cracking crude oil directly into ethylene, whereby heavy oil fractions and oil residues (bunker oil) can be used as feedstock instead of naphtha. In addition to processes developed by BASF some years ago, a new process has been jointly developed by Chiyoda Chemical Engineering and Kureha Chemical Industry (Japan). This process makes it possible for even sour crude oil to be used for making ethylene directly. An agreement on further development and commercialization of this process was signed with Union Carbide Corporation (UCC), USA, in December 1973. UCC was to build and test a prototype unit and commence the construction of a fully commercial-scale plant (Union Carbide has experience in separation and purification of ethylene, and also in pilot tests of crude oil cracking).[2]

Among other developments related to the utilization of heavy oil is the Flexicoker process developed by Esso Research & Engineering. Toa Oil Co. of Japan is reported to be installing a 21,000 barrels-per-stream-

1. Wolpert, *Synthetic and Plastics Paper*, Vol. 4, pp. 810-12.

2. Ibid., pp. 781, 809.

daily plant, utilizing this Esso process in its Kawasaki refinery. It is expected that the rate of recovering liquid fuel from heavy oil will be about 80%.

Plastics Products Production Costs

The important aspect of the production cost of synthetic polymer products is that the cost factor of crude oil is diluted in the various consecutive production stages from crude oil to finished products. As Shell has pointed out, "the cost of the chemical components in plastic bottles or a suit of clothes is comparatively small, and it should be possible to achieve the necessary improvement in price levels of basic chemicals without seriously damaging the overall demand for them."

A recent ICI publication stresses the point that although since autumn of 1973 the prices of traditional materials with which the major ICI thermoplastics compete have risen less than those of thermoplastics, it has been found that the increases in price for the finished (i.e., fabricated) plastics are, in most instances, of the same order as those of competing finished products made from other materials.[3] This is partly because of the lower density of plastics, which results in a greater volume of finished product being produced per ton of raw material, and partly because the manufacture of finished products in plastics is generally less labor intensive than manufacture in other materials; therefore, higher-wage costs have a smaller effect. These factors are cumulative, and the effect of the increased cost of oil diminishes with each processing step. The impact of increased oil prices in downstream operations, mid-1974 as against mid-1973, is shown in Table 6.3.

ENERGY CONSUMPTION

On the question of energy consumption the ICI publication states that "to make articles from plastics requires, in general, less energy than to make them from traditional materials, even allowing for the oil used for feedstock."[4]

The publication includes a list of U.K. energy requirements in the production of basic materials expressed in total TOE (metric ton of oil equivalent) per unit weight of product or material, and in kilocalories (kcal) per unit volume (taken as the cumulative total of direct and indirect energy consumed in successive manufacturing stages). The list

3. *The Competitiveness of Low Density Polyethylene, Polypropylene and Polyvinyl Chloride After the 1973 Oil Crisis: The ICI View* (London: ICI, 1974).

4. Ibid.

Table 6.3 *Effect of Increased Oil Prices (% Increase)*

Crude oil	+300	+300
Ethylene	+200	
Propylene		+180
Polyethylene	+100	
Polypropylene		+ 70
PE film	+ 50	
PE film bags	+ 30	
PP mouldings		+ 35

Source: Compiled from ICI, *The Competitiveness of Low Density Polyethylene, Polypropylene and Polyvinyl Chloride After the 1973 Oil Crisis: The ICI View* (London: ICI, 1974).

Note: Naphtha is not included in this table because it is subject to other demand pressures and can fluctuate independently of crude oil prices. The mid-1974 price shows an increase of more than 400% over the price of early 1973.

shows that energy consumption in the production of paper and board is slightly higher than in the production of polyethylene and polypropylene. However, if one adds the actual feedstock requirements in the production of plastics materials, then according to the ICI list the total energy consumption, expressed in kcal/cm^3, in the case of PE and PP is about 100% higher than in the case of paper and board. It is significant that in the case of finished (fabricated) products the situation is more favorable for plastics.

The U.K. energy requirements of oil (expressed as MT of oil equivalent) required as feedstock *and* to provide energy for manufacture of 1 million fertilizer sacks is 470 in the case of polyethylene sacks, and 700 in the case of paper sacks. The energy consumption in the production of cellulose film is nearly three times the total energy consumption in the production of HD PE and PP; in the production of 1 million m^2 of packaging film the total energy requirements are 110 in the case of PP film, and 155 in the case of cellulose film.

Other investigations of energy consumption in the production of various materials and manufactured products have resulted in the conclusion that the energy required for plastics is less than the energy required for paper and glass.

Table 6.4 shows energy consumption/MT of polyethylene bags (from distillation of naphtha from crude oil to extrusion of bags from PE). If one adds the required amount of crude oil (since it is a form of potential energy), then the total energy consumption is 96 x 109 Joules/MT,

Table 6.4 *Energy Consumption of Polyethylene Bags*

	10^9 *Joules/MT*
Distillation of naphtha from crude oil	1.23
Production of ethylene from naphtha	15.80
Polymerization of ethylene to polyethylene	10.55
Extrusion of bags from polyethylene	3.42
Total	31.0

Source: G. Schuur, Shell, article in *Verpakking*, no. 7 (1972), pp. 424-30, quoted in *Energy Conservation: Ways and Means*, p. 48.

including the conversion process energy consumption of 31 x 109 Joules/MT.[5] Within the conversion process, the production of ethylene and the polymerization of ethylene to polyethylene account, as the table shows, for 26.35 out of the total of 31 x 109 Joules/MT. It is possible that crude oil will someday be replaced as an energy source and will be used only as feedstock.

Reference is to be made to a joint high temperature gas-cooled reactor (HTGR) multi-unit project by Dow Chemical Complex and the Consumers Power in Midland (Michigan), and generally to the possibility of using HTGR in petroleum refining and petrochemical plants (possibly in a process of making ethylene directly from crude oil instead of from naphtha).[6]

According to Schuur 1.41 tons of crude oil are required to produce 1.38 tons of naphtha, 1.05 tons of ethylene, and 1.00 ton of LD PE, whereby the difference of 0.41 tons (1.41 minus 1.00) partly consists of by-products which are converted into steam or electric energy in the refinery or chemical plant. Approximately 2 tons of crude oil are required (feedstock plus energy source) to produce 1 ton of polymer, and this low consumption of crude oil is due to the fact that plastics are processed at low temperatures, the cooling-water consumption is low. The energy consumption is usually of the order of 0.15–0.30 kwh per kg of polymer.[7]

5. *Energy Conservation: Ways and Means*, ed. J.A. Over, (The Hague: Future Shape of Technology Foundation, 1974), p. 48. Chapter 3 by the Working Group "Industrial Energy Use."

6. T.D. Anderson and I. Spiewak, Oak Ridge National Laboratory, Tennessee, USA, "Study on the Applications of HTGR to a Petroleum Refinery-Petrochemical Concern," Paper in *The High Temperature Reactor and Process Applications Conference Proceedings* (London: The British Nuclear Energy Society, 1974).

7. Wolpert, *Synthetic and Plastics Paper*, Vol. 4, pp. 792-3, quoting G. Schuur's lecture "Environment, Raw Materials and Plastics" in The Hague, November 16, 1972.

Table 6.5 *Typical Energy Contents*

	Energy (megajoules/kg)	*Cost of energy/ value of product*
Plastics	10	0.04
Paper	25	0.3
Glass (bottles)	30-50	0.3

Source: Compiled from NATO, *Technology of Efficient Energy Utilization* (Report of the Science Committee Conference in October 1973), Brussels, p. 56.

Note: These are typical values. The actual value depends on purity, form, manufacturing process, and other variables.

NATO Evaluation

According to a NATO report the typical energy contents of materials and manufactured products compare as shown in Table 6.5. This report contains the following reference to the energy requirements in the drying operation during the paper production:

> The present technique for making paper involves the formation of a pulp slurry, the water content of which is reduced from 99.5% to a final, ideal value of 7%. The water is removed mechanically down to about 70%; from then on the removal is by evaporation at low efficiency because in current practice this is done without heat recycling. *Even ideally, the current drying operation requires a minimum of 1.5 MWh/tonne of paper, but in practice at least three times this energy is consumed.* Alternative, non-wet methods are required. Although these would involve drying the pulp in order to produce a fibre feedstock, they could be carried out at very low power requirements using a heat pump (e.g. with 10% of the energy currently used). Electrostatic laying of dry fibre has been attempted and has been found technically feasible at low production rates. Other adhesive techniques may be possible.
>
> This is an area requiring a good deal of technological research. However, in view of the fact that the paper industry consumes a not negligible fraction of the total energy (1.5% in the U.S.A.) this effort is justified. Capital investment as well as technical considerations make this a medium to long term project [italics mine].[8]

8. NATO, *Technology of Efficient Energy Utilization* (Report of the Science Committee Conference in October 1973), Brussels, ed. Eugene G. Kovach.

Midwest Research Institute Evaluation

A research project prepared for Mobil Chemical Company, Plastics Division, analyzed the *total environmental impact* of polystyrene foam trays and molded pulp trays, as against the more commonly used comparisons of individual sectors (e.g., disposal of waste).[9] The results of this research project showed that, at that point of time, polystyrene foam trays had lower overall environmental impact than molded pulp trays. The report warned against attempting to generalize from the specific conclusions in this particular research to overall conclusions that plastics are "better" than paper, etc., and added that "such judgments can be made only after specific product analysis."[10]

University of California Memorandum

A 1971 memorandum assessing energy and materials utilization in the USA (1968 data) listed the energy content of various consumer goods per unit as follows:[11,12]

Consumer goods	*kwht*
Plastics (packaging and containers)	~ 3.5 x 10^3/ton
Paper (average, including paperboard)	6,400/ton

This memorandum also compared the energy consumption (energy for unit production) in basic materials processing:

Material	*kwht/ton (2,000 lb)*
Plastics	2,400
Paper (finished, average)	5,900
Glass (plate finished)	6,700

9. Midwest Research Institute, "Environmental Impacts of Polystyrene Foam and Molded Pulp Meat Trays" (Kansas City, Missouri, USA, April 1972).

10. See Wolpert, *Synthetic and Plastics Paper*, Vol. 4, pp. 803-04.

11. Compiled from A.B. Makhijani and A.J. Lichtenberg, Electronics Research Laboratory, College of Engineering, University of California, Berkeley, "An Assessment of Energy and Materials Utilization in the U.S.A." Memorandum No. ERL-M310 (revised), September 1971.

12. Kwht = kilowatt hours thermal; where electricity is used in the production process, kilowatt hours electricity (kwhe) are converted to kwht by the formula kwht = kwhe/0.3. This includes generation as well as transmission losses.

Energy Consumption in Paper Production

In comparing the various energy-consumption estimates for the production of paper, it is necessary to take into account the differences in various countries, as well as whether both the energy needed for making pulp and the energy consumed in the now mechanized forestry industry are taken into account. For countries that depend on imports of pulp and produce only a small amount of pulp, or no pulp at all, the energy consumption involved in forestry and pulpmaking is only of interest in relation to how far these energy requirements find a reflection in the prices of imported pulp. In the Dutch paper industry the direct energy investment can be estimated at 18.8 x 10^9 Joules/MT.[13] Wood, the raw material, is itself a form of potential energy, which is lost when it is converted into packaging material or printing paper. This energy may re-emerge for use if the waste packaging is burnt and the heat is used to generate electricity. If this potential energy is included, then the energy consumption figure goes up from 18.8 x 10^9 Joules/MT to an average of 50 x 10^9 Joules/MT of packaging material (i.e., paper and cardboard).

According to W.P. Fornerod the energy costs in the West German paper industry are about 8.1% of the gross production value.[14] The Feldmühle 1973 Annual Report said that higher energy costs have seriously hit the paper industry, one of the most energy-intensive users. During the year the cost of heavy fuel oil alone more than doubled, yet for technical reasons the substitution of fuel oil by other sources of energy was possible only to a very limited extent.

It is noteworthy that the study "Energy Conservation: Ways and Means," which attempted as a first priority to concentrate on ways of easing the pressure on oil and gas, stated in the chapter on total consumption of energy for packaging materials that in some cases wood can be disregarded as a potential energy carrier. Although these considerations—from the point of a country's national economy—are fully justified, they may lead to importing certain materials and products (with high-energy content) instead of manufacturing them in the country concerned (e.g., the import of finished products like paper and plastics film for packaging, which would transfer the energy consumption from the importing country to the exporting country). The importing country would thus have correspondingly higher import bills and less employment for their own labor force.

During the last few years the pulp/paper industry has been making efforts to reduce its energy consumption, but it is still a high-energy-

13. *Energy Conservation: Ways and Means*, 1974.

14. From Packaging TNO Report No. 14/73, quoted in *Energy Conservation: Ways and Means*, 1974, p. 48.

consuming industry. According to *Fuel Survey,* in 1974 in the USA, pulpwood harvesting alone required an average of 5.16 gallons of fuel—either gasoline or diesel—to harvest one cord of pulpwood and move it from the stump to a concentration point or mill by truck. *Fuel Survey* adds that "it appears doubtful that the average rate of fuel consumption for harvesting can be reduced" due to the trend toward greater mechanization.[15] According to a recent statement (October 31, 1975) by the International Paper Company (USA) it takes at least three barrels of purchased oil to make a ton of paper.

ENVIRONMENTAL IMPACT

Environmental impact must be evaluated in its total context, instead of being related (as is the case in many discussions) only to disposal of waste. Even in this single sector of the total problem, confusion is created by lumping together the question of *litter,* which is a purely social problem, and that of *waste disposal* (and recycling of waste), which is a techno-economic problem.

There are two main factors associated with the consumption and production of products:

1. Utilization of the earth's natural resources (raw materials, energy, and water).
2. The environmental cost attached to these products' production and disposal, in the form of solid waste, air emissions, and water effluents.

The individual components of these two overall factors have to be analyzed in order to be able to assess their total environmental impact. Dr. W.C. Fergusson, I.C.I., England, in his presentation "Plastics and the Environment—U.K. National Situation Report" at the 1973 ANTEC Conference in Montreal, pointed out that

> a total appreciation must be made of the effect on energy and raw materials resources, of the gaining of the raw material and its effect on the environment, the conversion of raw material to the goods required, the amount of waste generated and its effect on the environment, and the balance between resources used in waste recovery and the real value of the recovered material. From this it should be possible to determine the best raw material and the total

15. Published by the American Pulpwood Association, Washington, D.C., 1975.

system of use involved for a particular purpose in the light of a true conservation policy.

Dr. Fergusson stressed the point that the use of plastics in packaging can result in the reduction of weight and volume of wastes, as well as in savings of raw material, and should be studied in this way. He concluded by saying that "plastics make not only a major contribution to the industrial welfare of our nation but to resource conservation and the protection of the environment."

At the ANTEC Montreal Conference two sessions were devoted to "Plastics and the Environment." Bernard Baum and Charles H. Parker of DeBell & Richardson, Inc., (USA) in their joint paper "Plastics Disposability and Recycling," referred to a number of leading firms that were either using or studying the possibility of reusing plastics in production of various products. Dow Chemical Company research workers have described a method whereby heterogeneous mixtures of plastics (separated from other municipal refuse) may be reworked into useful materials by being treated with up to 25% chlorinated polyethylene. Baum and Parker felt that the reuse of plastic products from the municipal waste stream requires further research and development before such reuse can become a practical method of disposal. For plastics, better recovery of energy or useful products may be more likely through pyrolysis or incineration with steam recovery.

The late Dr. E.F.J. Tomalin, BP Chemicals International Ltd., London, in his presentation "Designing for Disposability," said that in Great Britain the presence of plastics, notably as packaging materials, did not introduce any specific difficulties into the collection and disposal of waste, and that this situation was unlikely to change in the next 10 years. He felt that it was probably safe to assume that the average plastics content of the collected municipal waste in Britain was 2%. It was expected that this figure might rise to 5% by 1980. He added that PVC, which had particularly been under attack, represented less than 0.2% by weight of collected waste.

Albert J. Klee, Solid Waste Research Laboratory, U.S. Environmental Protection Agency, in his presentation "Environmental Plastics Legislation in the U.S.: An Overview," said that "among the literally hundreds of bills proposed involving plastics in an environmental context, almost all deal with the litter issue. . . ." He referred to the court case at the Supreme Court in New York, during which the court established that "it costs no more, and probably less, to collect plastic containers than to collect paper, metal or glass containers, as the significant cost factor in the municipal collection of solid waste is the weight of the refuse load . . . ," and that "expert testimony established

Table 6.6 *Composition of Roadside Litter in the USA*

Material	*Totals (%)*
Papers and periodicals	1.9
Paper packagings and boxes	11.5
Other paper articles	46.1
Tins	16.3
Plastics objects	5.8
Glass bottles and jars	5.9
Other objects	12.5

Source: Wolpert, *Synthetic and Plastics Paper,* Vol. 4, 1974, p. 792, (quoting from G. Schuur's lecture "Environment, Raw Materials, and Plastics," in The Hague, November 1972, which referred to these data from the "National Study on Roadside Litter," issued by the National Research Council, Highway Research Board, USA).

that plastic containers are cheaper to dispose of by incineration than glass or metal containers. These last-named types do not burn and some 30% of the cost of incineration is removal of the residue left after incineration. . . ."

At the Plastics and Paper Conference, London, May 1973, K.H.C. Bessant, BP Chemicals, in his presentation "Some Environmental Aspects of Plastics," said that petrochemical complexes have their own particular environmental problems, generally local ones, relating to noise, light (from the flares of burning-off gases), and possibly odor.

In general, polymerization plants do not present serious or large-scale environmental or effluent problems. The monetary cost of disposing of municipal refuse is primarily the cost of collecting, handling, and transporting it, and the average cost of collection and disposal of U.K. refuse was about £5/ton.[16] Plastics are generally welcome in modern refuse incineration plants, since they have low water absorption and help to burn up refuse, having a high calorific value. In his concluding remarks Bessant stressed the point that, even in such an apparently clear-cut case as returnable versus nonreturnable containers, it is by no means certain that resource conservation is always best served by adopting the returnables. Table 6.6 lists the composition of roadside litter in the USA.

16. Department of the Environment, "Refuse Disposal Report" (London: H.M.S.O., 1971); and "Interim Report of the Working Party on Design for Disposability" (London: Plastics Institute, 1972). See also Wolpert, *Synthetic and Plastics Paper*, Vol. 4, pp. 791-92.

Pulp and Paper Industry—Water- and Air-Pollution Abatement

J.S. Evans, British Paper and Board Makers' Association, in his presentation "The Pulp and Paper Industry and the Environment" at the Plastics and Paper Conference, made the following points:

1. An abundance of good water has always been a manufacturing necessity for the industry, and water-courses have served for the discharge of effluent.
2. The total water used by the pulp and paper industries of all OECD countries was estimated at 21,842 million cubic meters in 1970.[17]
3. It is possible to considerably reduce the water use in the design of modern equipment.

On the question of effluents from the pulp and paper industry, Evans remarked that "the normal sewage method of treatment of paper-mill effluent is a practical proposition. The only obstacle is one of cost." He referred to two fundamentally different philosophical approaches underlying certain legislations; namely, in some countries the basic philosophy is to maintain or restore the natural quality of water, whereas in other countries the basic philosophy is a compromise between the various uses that man makes of the water, including industrial development and the disposal of waste. Some governments intervene in order to finance this expenditure in different ways (these types of support of the pulp/paper industry are, of course, subsidies paid by the tax payers of those countries).

Evans felt that during the next few years investment in pollution control would probably account for a substantial part of total investment in the pulp and paper industry, and he stressed the point that "it would appear that because of its small profit margins, the pulp and paper industry may find it difficult to obtain the necessary financial resources, even for productive investment." In Britain, according to Evans, the paper and board makers invested £6.5 million from 1965 to 1970 in capital equipment for effluent plant, and a further £6.2 million for running expenses. During 1971-72 an additional £1.8 million was spent on capital expenditures. It was also estimated that the improved standards will require the expenditure of a further £17.5 million between 1973 and 1977.

17. According to *Pollution by the Pulp and Paper Industry* (Paris: OECD, autumn 1973), the water consumption of the pulp and paper industry in some countries is responsible for 70% of the total industrial and domestic water withdrawal (excluding cooling water for power stations). A typical sulfate mill producing 500 MT/day uses about 100,000 m^3 of water daily, an equivalent to the water consumption of a city with a population of half a million.

The 1972 OECD Report "Advanced Pollution Abatement Technology in the Pulp and Paper Industry" said in its conclusion that[18]

1. In the case of water pollution the costs for effluent treatment vary from country to country, but as an indication it can be approximated that a reduction of 90% of the waste load from the effluent of a bleached kraft mill causes an extra cost of $7–8/ton of pulp.
2. Regarding air-pollution abatement (odorous gaseous emissions from kraft mills are the main source of air pollution), in the past few years new kraft mills have been constructed that meet the highest established standards of air-pollution control. The report adds, however, that efficient, presently available techniques are very expensive, and that the cost of air pollution control might eventually be equivalent to that of water-pollution control in the kraft industry.

This means that the total cost of water- and air-pollution control may be as high as $14–16/ton.

There is a great need for the development of cost-control techniques. The report says that a great emphasis is being placed on the development of processes that do not use sulphur- or chlorine-containing chemicals. Improved mechanical pulps may have an important environmental impact by eliminating or reducing the need for chemical furnish in newsprint manufacture. However, the report states that none of the new approaches to pulping will receive wide commercial acceptability before 1980. Their impact should be felt within 15 to 20 years (i.e., by the end of the eighties or beginning of the nineties).

On a slightly more optimistic note, the report concluded that the industry recognized the need for reducing the pollution load from pulp-bleaching systems and was developing promising new process technology in this area. The indications were that new bleaching technology would soon become available that would permit radical reductions in mill water requirements and in the pollution load. Also, important R&D expenditures would have to be made over the next few years, if these, or still undiscovered, technologies were to have a significant impact on the industry's pollution-control efforts. In the OECD publication *Pollution by the Pulp and Paper Industry (Present Situation and Trends),* it was stated that as far as the paper industry is concerned, pollution-control costs, including those related to chemical furnish, may affect the competitiveness of paper with other products.[19] This is particularly true for

18. This report, by a group of experts, was adopted in June 1972 by the joint ad hoc group of the OECD Environment Committee on Pollution by the Pulp and Paper Industry, and published by the OECD in Paris, February 1973.

19. Chapter entitled "Commentary on Economic and Financial Matters."

wrapping and packaging papers, since these grades can, to a certain extent, be replaced by plastics.

It is expected that structural changes within the pulp and paper industry will be accelerated as a result of pollution-abatement costs. Small, less profitable mills will be closed, and the concentration of production in large units will be accelerated. Pollution-control costs for sulfite pulp are expected to be twice as high as those for sulfate pulp. Since the profitability of sulfite-pulp production and its competitiveness with other grades have been unsatisfactory during the last few years, it is to be expected that a further deterioration in the competitiveness and profitability of sulfite mills will occur.

The semichemical pulp industry is also expected to experience high pollution-control costs, and the relatively high pollution-control costs may cause the competitiveness of this sector of the pulp industry to decline. According to this OECD report the estimated pollution-control costs were to increase between 1970–75–80 as follows.[20]

Semichemical Pulp

The average increase of control costs was to be from \$2.53 in 1970 to \$12.05 in 1975 (per MT). In the case of Canada the increase was to be from \$0.25 in 1970 to \$6.03 in 1975, and to \$8.78 in 1980. These costs would have averaged 2.0% in 1970 and 9.6% in 1975 of product price.

Sulfite Pulp

The average increase of control costs was to be from \$1.22 in 1970 to \$13.53 in 1975 (per MT). Whereas in 1970 these costs accounted for 1.9% of product price, they were expected to rise to 7.8% in 1975 (in the case of Canada they are expected to account for 10.0%, in the case of France to 13.5%, and in the case of Sweden to 9.2% of product price in 1980).

Nonintegrated Sulfate Pulp

The average increase of control costs was to be from \$1.22 in 1970 to \$4.55 in 1975 (in 1980 it is expected to reach \$8.20/MT in Sweden).

Integrated Sulfate Pulp and Paper

The average increase of control costs was expected to increase from \$1.88 in 1970 to \$10.05 in 1975 (per MT). It was estimated that in 1980

20. These figures are based on the December 1970 value and exchange rates of the U.S. dollar. In calculating the percentage of product price, the following selling prices have been used as a basis: semichemical pulp, \$125/MT; sulfite pulp, \$173/MT; nonintegrated sulfate pulp, \$177/MT; integrated sulfate pulp, \$173/MT.

Table 6.7 *Estimated Pollution-Control Costs in 1980 in the USA*

Product category	*Production (thousand MT)*	*Pollution control costs ($/MT)*		
		Water	*Air*	*Total*
Semichemical pulp	3,035	12.40	6.00	18.40
Sulphite pulp	2,968	18.20	5.80	24.00
Integrated sulphate pulp and paper	26,798	10.30	9.90	21.20
Newsprint	3,000	6.10	–	6.10
Other paper and board	19,000	6.20	1.30	7.50

Source: OECD, *Pollution by the Pulp and Paper Industry* (Paris: OECD, autumn 1973).

Note: These cost figures were tabled at the last meeting of the ad hoc group in February, 1973; they are based on the consideration of legislation enacted in late 1972, i.e., after completion of pollution control cost surveys in all other participating countries.

these control costs would amount to $6.57 in Canada, $7.04 in France, and $7.38 in Sweden. In the USA the estimated increase was to be from $2.21 in 1970 to $11.80 in 1975.

Newsprint

In Canada, where pollution-control costs for newsprint include pollution-control costs incurred in the manufacture of chemical pulp used in newsprint (approximately 20% of total fiber in newsprint is chemical pulp), it was estimated that the pollution-control costs would go up from $0.51 in 1970 to $3.55 in 1975 and to $6.55 in 1980 (per MT). In other countries pollution-control costs for newsprint do not include pollution-control costs incurred in manufacture of chemical pulp used in newsprint, and are therefore lower than those in Canada. In the USA they were expected to increase from $0.91 in 1970 to $3.75 in 1975, and in Sweden from $3.62 in 1970 to $4.86 in 1975, and then to decrease to $4.01 in 1980.

Other Paper and Board

The average increase of control costs was estimated to increase from $1.11 in 1970 to $3.80 in 1975 (per MT). In Canada the pollution-control costs, at $0.13 in 1970, were estimated to increase to $3.23 in 1975 and $3.69 in 1980. In Fed Rep Germany the pollution-control costs were estimated to increase from $3.14 in 1970 to $3.52 by 1975 and to $5.41 in 1980.

Table 6.7 estimates the pollution-control costs in the USA in 1980;

these estimates take into account legislation enacted in late 1972.

The cost of R&D work in this field is already contributing to the production costs of pulp and paper. R&D costs on pollution abatement in 1971 in the pulp and paper industry expressed in US$/ton of production varied from $0.09 in Norway and Netherlands, to $0.24 in Britain, to $0.28 in Canada, and to $0.34 in the USA. In the USA the total R&D expenditure in this industry for pollution abatement amounted to $19.0 million in 1971.

The OECD report says in conclusion that

> Until recently, it has been usual for the pulp and paper industry to fund the total cost of controlling pollution. Since the coming into force of stricter regulations in most member countries, many governments now intervene in some way to subsidise directly or indirectly this expenditure. . . . The extent of such subsidisation also varies among countries, and such variations will contribute to the overall effect on competitiveness. . . . Pollution control costs will account for a substantial part of total capital investments for old mills in the short term, and . . . due to its record of low profitability, the industry in many countries may find it difficult to obtain the necessary capital even for productive investments.

It is evident that these pollution-control measures, including the R&D expenditures, will be an additional factor contributing to increased production costs of the paper industry, resulting in higher prices of paper and paperboard.

OUTLOOK

One of the characteristic features of the paper industry's markets is that within the overall growing consumption (which is expected to continue to increase) the periods of recession and of growth are becoming shorter and more pronounced. This trend is affecting the industry's investments in new equipment, since the financial returns are very low during the repeated recession periods and these recession periods, together with the prevailing high interest rates and the heavily increased cost of new equipment, make it difficult to attract capital for new production units.

The trend towards vertical integration within the forest-owned pulp and paper industry results in a shortage of marketing pulp for papermaking firms that depend on buying pulp from outside sources. This shortage affects the nonintegrated papermaking firms particularly during the periods of strong demand for paper and paper products.[21]

According to the FAO document *World Pulp and Paper Capacities 1973 to 1978, FAO Survey 1974,* the world paper-pulp capacity was estimated at 133.8 million MT in 1974, and in the same year the market-pulp capacity (paper grades) was stated to be 25.8 million MT (including 23.5 million MT of chemical woodpulp). These figures clearly indicate the degree of vertical integration within the pulp-paper industry, and this trend of further integration is expected to continue. The world paper-pulp capacity is expected to grow at an average of 3.4%/yr and to reach 151.7 million MT in 1978, whereas the market-pulp capacity is expected to grow at a lower rate, reflecting the improved markets for paper and paperboard from integrated operations. Between 1974 and 1976, the total paper-pulp production capacity was expected to increase by 9.7 million MT, or by over 7%, whereas the market-pulp capacity was expected to increase by 1.5 million MT, or by less than 6%.

The other important point is that the ratio between the paper-pulp and the paper and paperboard capacities now shows a declining trend, namely from 79.8% in 1973 and 1974 to 78.8% in 1975; the estimated ratio for the 1976-78 period is 78.6%. This trend is contrary to the developments foreseen in the three previous FAO surveys, which indicated an increasing ratio; the change in trend is the result of two main factors:

1. The increased paper and paperboard and decreased paper-pulp capacity estimates in Japan.
2. Paper and paperboard capacity estimates are increasing more rapidly than paper-pulp capacity estimates in North America and Western Europe.

According to the FAO document, the rated capacity of the paper and board industry on a global basis is expected to increase as follows:

Year	*Million MT*
1973	160.6
1974	167.8
1975	175.9
1976	182.5
1977	188.5
1978	192.9

21. In connection with paper shortages during boom periods, it is significant that mergers or financial participation between newspaper or periodical publishers and pulp/papermaking firms have recently taken place in Fed Rep Germany. In Britain, Reed International, a papermaking company, has very widespread publishing activities.

Table 6.8 *Growth by Main Types of Paper and Board Capacities – 1973-78*

	Average annual growth rate (%)	*Share of total paper and paperboard capacity (%)*
Newsprint	3.1	15.0
Printing and writing paper	3.7	22.0
Other papers and paperboard*	almost 4.0	63.0

Source: FAO, *World Pulp and Paper Capacities 1973 to 1978, FAO Survey 1974* (Rome 1974).

*Mainly comprising packaging grades.

The average annual growth rate during the 1973-78 period is forecast to amount to 3.7%, as against 5.8% during the 1963-68 period and 4.7% during the 1968-73 period. But the FAO document adds that the capacity forecasts, especially for the last two years of the 5-year forecasting period, tend to be too low and consequently reduce the 5-year average growth rate. The breakdown by the main types of paper and board capacities during the 1973-78 period is estimated in Table 6.8. The table gives another indication of the growing importance of the packaging field for the paper and paperboard industry, a field where competition and cooperation between paper and synthetic polymer products are stronger than in other fields (apart from the use of synthetic polymers in the production of printing and writing paper).

This FAO document does not include any forecasts on the consumption growth of paper and paperboard, nor on the factor of capital inflow constraint on production in the market economies. This factor was introduced by the FAO in its revised document *Outlook for Pulp and Paper Consumption, Production and Trade* in March 1972.[22] In that document, the FAO felt that the consumption outlook should reflect a capital inflow constraint on production in the market economies. Table 6.9 compares the outlook for consumption of paper and a production constraint outlook for consumption of paper and board.

These considerations include the question of the prospective financial rate of return on capital of the market economies in the pulp and paper industry and the price level which would generate and attract sufficient funds to establish the additional pulp and paper production capacity required to reach the forecast consumption level. The 1972 FAO report said that whereas the indications were that the demand for paper has been quite inelastic, the industry was not confident that the

22. See Wolpert, *Synthetic and Plastics Paper*, Vol. 3, pp. 371-75.

Table 6.9 *Consumption and Production Outlooks (in Million MT)*

	1975	*1980*	*1985*
Outlook for consumption of paper	168.60	218.2	281.63
Production constrained outlook for consumption	159.6	207.0	260.0

Source: FAO, *Outlook for Pulp and Paper Consumption, Production and Trade* (F/PAP/DST/71/1.1/Rev.), March 1972, quoted in Wolpert, *Synthetic and Plastics Paper*, Vol. 3, 1972, pp. 371-75.

Note: Figures refer to the total of paper and paperboard.

demand would stay this way. The report also said that the widening range and competitiveness of substitutes gave some support to the possibility that demand for paper will become more elastic.

In its 1973 Annual Report, Weyerhaeuser referred to the pulp business being increasingly tied to waste utilization, using wood-products manufacturing and forest wastes, including sawdust. On the supply-demand position of pulp the report said that

> pulp recently was in commodity surplus. Demand began to catch up with supply in 1972, and passed it last year, with the world shortage expected to become increasingly severe through the 1970s. Unless considerable industry investment flows into pulping within the next two to three years, world demand could exceed supply by several million tons by 1980. Because of environmental restrictions, and poor past markets, many mills have actually closed, or are faced with closure.

With reference to products from bleached pulp, the report said that "bleached paperboards have already moved into short supply, and the shortage is expected to intensify as rapidly as for other bleached pulp products."

On the use of bleached pulp by other than papermaking and boardmaking industries, the report said that bleached pulps are also used for "chemical conversion into hundreds of other items ranging from textiles to plastics." This is another indication of cooperation between the pulp and the synthetic polymer industries.

Some industrialists estimate that there will be a chronic shortage of wood pulp coming from traditional sources, which will amount to about 3% of the total pulp production needed to satisfy the paper industry's requirements—that is, about 4 million MT/yr growing to over 4.5 million MT/yr by the end of this decade. It is felt that this shortage will be partly alleviated by new technologies, including those making a greater use of the full tree utilization approach. The remaining shortage will not be covered by a dramatic upsurge of a single new raw material, but by the sum of several materials. A greater use of nonwood fibers, for example, bagasse (at present only 1 million MT of bagasse out of the available annual amount of about 20 million MT is being used for pulping purposes), bamboo, reed, and possibly straw, will be made. These nonwood indigenous fibers when used in certain developing countries will have the effect of relieving to some extent the pressure on the supply position of wood fibers from traditional sources.

Seen in this context, synthetic pulp will become an additional source of raw material for the paper industry in the 1980s, starting as an additive to conventional pulp; later, it is possible that synthetic paper made from 100% synthetic pulp will be used for special applications, as indicated in Chapter 4.

Price Competitiveness

Since the crude oil price increases in September 1973, the following factors have influenced the competitive position between plastics and paper, and in some countries government controls related to prices and their increases have been distorting the influence of the market factors.

Polymers and plastics, the production costs of which went up as a result of the increase in crude oil prices, were in very short supply during the second half of 1973 and at the beginning of 1974. These shortages, which occurred at the time of a high level of economic activity in the industrialized countries with a buoyant demand for petrochemicals, were partly due to the speculative buying in world markets for feedstocks, intermediates, and polymers, the hoarding of these materials (reflecting the expectancy of higher prices), and the growing consumption of these materials (resulting in "black market" prices). By the end of 1974, however, the shortage gave way to more normal conditions. Plastics fabricators who had purchased larger quantities of polymers than they needed for their short-term requirements began using up their stocks, with the result that polymer manufacturers were able to abandon rationing and were able to come again into a position of supplying to their clients at least some polymers on short-term notice or sometimes even from stock. This coincided with the overall recession which began toward the end of 1974, and which included the recession in the motorcar industry and in the building trade; by the end of 1974 the prices of polymers were as much as 10% lower than at the beginning of that year.

Throughout 1975 and the beginning of 1976 the recession continued in Western Europe, but by the end of 1975 certain price increases, reflecting increased production costs, were registered. In the USA the economic recovery had already begun by the end of 1975, and price increases occurred. In both cases the price level reached the early 1974 level, in some cases registering increases within the overall inflation rate.

Paper prices, which in 1971-72 reflected the recession in this industry and were uneconomically low, increased with the improvement of the market in 1973. Considerable increases in pulp prices, energy costs, etc., plus the shortage of market pulp (and in some cases of paper and paper products in various countries) have contributed to the increase of prices of paper. In Europe the paper prices increased by about 50% between mid-1973 and mid-1974 (the actual increased production costs are reported to have been even higher than the increase of paper prices), and further increases took place after mid-1974. This trend is expected to continue.

A recession in the paper trade which began in Europe at the end of 1974 and continued through 1975 is expected to end by the end of 1976. Some leading German paper manufacturers felt that in the case of an accentuated recession it was more advisable to put some of their machines out of commission temporarily than to embark in cutthroat competition (because of over-supply) that would result in low prices and annual financial losses, as occurred during the previous recession.

Despite this intention paper and board prices declined during the 1975 recession. It is to be expected that during the forthcoming recovery the prices of paper and board will not only regain the 1974 level but will have to be increased even above the overall inflation rate. In its statement to its shareholders on October 31, 1975, the International Paper Company (USA) said that "Inflation and increasing costs have deflated the real worth of price increases, so that again return on investments for most product lines does not warrant the high costs and high risks of new capacity."

The question of securing (or earning) the required funds for expansion of production capacities in order to be able to meet the overall increasing consumption of paper and paperboard is a highly acute problem for the paper industry. Dr. H. Gorsler, Papierfabrik Albbruck (Fed Rep Germany), speaking at the International Pulp and Paper Conference in Stockholm (October 1974), stated that although the immediate consequences of the increased oil prices had led to high increases of prices of plastics and to a shortage of plastics products, including carrier bags—resulting in an increased use of paper bags—the situation had returned to what it was before the start of the oil crisis.

The paper industry is justified in looking toward a growing demand for its products (despite the downs during the regularly occurring recession periods). Large-scale investments by the pulp and paper industry in order to meet these increasing demands can, however, only be secured by a higher profitability of this industry, which in turn can only be achieved by paper prices increasing in excess of the overall inflation rate.

The confidence of the polymer industry in the competitive strength of its products was clearly demonstrated by the high investments planned for large-scale increases of production capacities in 1975-76 by leading firms of this industry.

In the synthetic polymer industry there is always competition between the various polymers, even between processes for making semimanufactured and end products. Synthetic papers of second generation based on plastics film, and synthetic papers based on synthetic pulp, will be competing for the same markets. At present, polypropylene and polyester, as well as other plastics films replacing paper products, have achieved a penetration of specialized markets in the electrical industry, including cable wrapping and condensers (capacitors) production. Yet it is possible that synthetic papers (based on 100% synthetic pulp or on synthetic plus conventional pulp) will secure a niche in these markets.

An interesting example of competition within the synthetic polymer industry has been the introduction of flat expanded PP and HD PE tapes (made from extruded film) for making carpet backing, sacks, and woolpacks in competition with jute and partly with kraft sacks. These tapes are woven, and in some cases knitted, to fabrics. This development was first ridiculed by synthetic fiber manufacturers applying the traditional spinnerette process. However, during the last eight years, products made from this type of flat plastics tape have secured a significant market penetration and have proved that the economics of this process are a viable proposition (despite the fact that polymers are first extruded into film and then slit into tapes, i.e., flat-yarn instead of direct-yarn production from polymers).

The next step was the fibrillation of these tapes into surface-carpet yarns, and there are further potential developments in this field. In the meantime PP spunbonded materials, including Typar, have been competing with woven-tape fabrics in carpet-backing markets, and Valeron special laminates of two HD PE sheets started to compete with woven-tape fabrics (sometimes used together with kraft paper) in some packaging applications.

It can be assumed that various types of synthetic paper (based on film *and* on synthetic pulp from various polymers) will achieve a growing penetration of the market. The choice of the polymer and of the process in question for individual applications will be decided by the required properties of the end product and the size of its market.

The acceptance of synthetic pulp as an additional material by papermaking firms will be a gradual one; the time that it took the industry to accept hardwood pulp is to be remembered in this context. Parallel to these developments, plastics paper (paper-like HD PE film) will continue to increase its share of the overall packaging market because of the functional properties of plastics paper.

The increasing shortage of market pulp during boom periods, combined with its increasing prices, will stimulate a growing number of papermaking firms to diversify their activities into the plastics field and to use synthetic pulp as an additional raw material. This, of course, will represent a further step in cooperation between the paper and the synthetic polymer industries.

BIBLIOGRAPHY

Allgemeine Papier-Rundschau, Fed Rep Germany, April 1972.

ANTEC Technical Conference, Montreal, Canada, May 1973. Organized by The Society of Plastics Engineers, Greenwich, Connecticut, USA.

Arthur D. Little, Inc. "Packaging in Perspective, Report to the Ad Hoc Committee on Packaging." Cambridge, Massachusetts, USA: February 1974.

BASF, Fed Rep Germany, Annual Reports, 1971 and 1972.

Briston, J. H. "Plastics Films." Published for The Plastics Institute. London: Iliffe Books Ltd., 1974.

The British Nuclear Energy Society. *The High Temperature Reactor and Process Applications Conference Proceedings*. London: November 1974.

BPBMA. "The Fundamental Properties of Paper Related to its Uses," International Symposium, Cambridge, England, September 1973. Organized by the Technical Section, The British Paper and Board Makers' Association, London, in association with Paper Physics Committee of TAPPI, and Research Committee, CPPA, Canada. Proceedings available from BPBMA.

Brydson, J. A. *Plastics Materials*. London: Iliffe Books Ltd., and Princeton, New Jersey, USA: D. Van Nostrand Company, 1966.

Chemicals EDC. 1972 Report of the Plastics Working Party, London.

DuPont Magazine, Geneva, May 1974.

Energy Conservation: Ways and Means. The Hague: Future Shape of Technology Foundation, 1974.

European Chemical Market Research Conference, October 1973.

FAO. "Outlook for Pulp and Paper Consumption, Production and Trade." FAO document, March 1972.

FAO. "World Pulp and Paper Capacities, 1973-1978, Survey, 1974." FAO document, 1974.

Feldmühle, Fed Rep Germany, Annual Reports, 1972 and 1973.

Fuji Bank Bulletin, Tokyo, May 1973.

Hercules, USA, Annual Report, 1972.

Hoge, Dr. William H. "Synpulp Projects & Technologies." Flemington, New Jersey, USA: Paper Associates, 1973.

ICI. *The Competitiveness of Low Density Polyethylene, Polypropylene and Polyvinyl Chloride after the 1973 Oil Crisis: The ICI View.* London: 1974.

IUPAC-EUCEPA. *Man-Made Polymers in Papermaking: Proceedings of the IUPAC-EUCEPA Symposium, Helsinki, June, 1972.* Helsinki: Finnish Pulp and Paper Research Institute, 1973.

Makhijani, A. B. and Lichtenberg, A. J. "An Assessment of Energy and Materials Utilization in the U.S.A." The Electronics Research Laboratory, College of Engineering, University of California, Berkeley, California, USA: Memorandum No. ERL-M310 (revised), September 1971.

Midwest Research Institute. "Environmental Impacts of Polystyrene Foam and Molded Pulp Meat Trays." Prepared for Mobil Chemical Company, Plastics Division. Kansas City, Missouri, USA: April 1972.

Mitchell, Hutchins Inc. *Status Report:* "The Paper Industry—Will It Over-Expand?" New York, USA: January 1974.

Monsanto, USA, Annual Reports, 1972 and 1973.

NATO. *Technology of Efficient Energy Utilization.* Edited by Eugene G. Kovach. Report of a Science Committee Conference, October 1973, Brussels.

OECD. "Advanced Pollution Abatement Technology in the Pulp and Paper Industry." Paris: February 1973.

———. *Pollution by the Pulp and Paper Industry* (present situation and trends). Paris: autumn 1973.

———. "The Pulp and Paper Industry," annual reports. Paris: 1972-1973 (1974); 1973-1974 (1975); 1974-1975 (1976).

Packaging Review, London, March 1973.

"Plastics and Paper" Conference, London, May 1973. Organized by The Plastics Institute, London, in association with the British Paper and Board Makers' Association, London. Conference preprints published by The Plastics Institute.

"Plastics Industries in a Developing World," Conference in London, June 1972. Organized by The Plastics Institute, London.

The Plastics Institute, *Proceedings of the Technical Conference, Brighton, England, April 30-May 1, 1970.*

Pulp & Paper International, Brussels, August 1974.

Reed International, UK, Annual Report, 1973.

St. Regis Paper Company, USA, Annual Reports, 1972 and 1974.

Shell Chemicals. "Petroleum-based Chemicals—the Feedstock Problem." London: October 1973.

Stanford Research Institute. "The Plastics Industry in the Year 2000." Report prepared for The Society of the Plastics Industry, Inc., and The Society of Plastics Engineers, Greenwich, Connecticut, USA. Menlo Park, California, USA: April 1973.

Synthetic Fibers in Papermaking. Edited by O. A. Battista, New York: John Wiley & Sons, Inc., Interscience Publishers, 1964.

Weyerhaeuser, USA, Annual Report, 1973.

Wolpert, Vladimir M. *Plastics Tapes for Textiles,* two volumes. Tenterden, Kent, England: Marketing Man-Made Materials, 1968.

———. *Shrink Film—A Growing Market*. London: Wolpert & Jones (Studies) Ltd., May 1973.

———. *Synthetic and Plastics Paper*, four-volume international techno-economic survey. London: Wolpert & Jones (Studies) Ltd., Vol. 1, February 1971; Vol. 2, January 1972; Vol. 3, November 1972; Vol. 4, February 1974.

———. *Plastics and the Paper Industry,* an international techno-economic survey. London: Wolpert & Jones (Studies) Ltd., August 1975.

INDEX

VLADIMIR M. WOLPERT

is an internationally known expert in the relationships of synthetic polymers to the paper industry. Born in Moscow and educated in Germany, he has lived in London since 1939, where he has been working as a research and marketing consultant and journalist. As director of Wolpert & Jones (Studies) Ltd., London, he has traveled extensively in Europe, North America, and the Far East. He thus brings solid international experience and knowledge to this important and informative book.

Wolpert is the author of a number of international techno-economic surveys, including *Plastics Tapes for Textiles*, (2 volumes, 1968), *Synthetic and Plastics Paper* (4 volumes, 1971-74), *Shrink Film—A Growing Market* (1973), and *Plastics and the Paper Industry* (2 volumes, 1975-76), and is co-author of *Composite Materials for Packaging* (1976). He has presented papers at various international conferences, including the IUPAC-EUCEPA symposium at Helsinki (1972), the ANTEC conference at Montreal (1973), the "Plastics and Paper" conference in London (1973), the TAPPI 60th Anniversary Meeting in New York (1975), the TAPPI International Symposium on Synthetic Papers and Pulps in Washington (1976), and the American Chemical Society Centennial Meeting in New York (1976). In addition, he has published numerous articles and reports in technical and business journals, including *Pulp & Paper International*.

A scanning electron micrograph of the surface of a sheet made with 100% SWP® synthetic pulp (magnification 1680x) has been reproduced on the cover of this book, courtesy of Crown Zellerbach, Central Research, Camas, Washington, USA.

Cover design by Debra Shishkoff.

University of Lowell
Libraries

University of Lowell
Libraries

PRINTED IN U.S.A.
BRODART CAT. NO. 23 012